Michael M. Dediu

Marvelously
Terraforming

around the Terra by the Year **3543**

Moving from looking around, to
terraforming everything by **3543**

Copyright ©2024   by Michael M. Dediu

All rights reserved

Published and printed in the United States of America

Library of Congress Control Number: 2024918251

Dediu, Michael M.

Marvelously Terraforming around the Terra by the Year 3543
Moving from looking around, to terraforming everything by 3543,

ISBN-13:

1589939_Um0h9rk5ggOyAnbjjhhh
1593804_HZ0kclZAoAl6cUl05Nix

Preface

Where are we and where are we going? We are at the beginning of an excellent advancement of our civilization, and we are going to have what all people wanted for millennia – peace everywhere (including in space!), no arms at all (what a joy!!), freedom & order (to quickly advance, without conflicts), good health (to fully enjoy life), good education (to be able to understand everything), good economy (including plenty of jobs, terraforming, etc.), harmony (to laugh and celebrate together), and prosperity for all (finally all happy!).

This book, using a planetary dialog, focuses on terraforming in the year 3543 – which means transforming other space entities (planets, moons, asteroids, etc.) to be as analogous as possible to Terra – and, also, presenting many new important ideas for Terra itself, remembering what Churchill, over 65 years ago, said:
"If the human race wishes to have a prolonged and indefinite period of material prosperity, they have only got to behave in a peaceful and helpful way toward one another".

The future begins to take shape, and looks grandiose!

Michael M. Dediu, Ph. D.

21 October 2024

USA, the University of California, Berkeley (1868, named after the philosopher and mathematician Bishop George Berkeley (1685-1753), motto Fiat lux (Let there be light), 36,200 students, major public research university, 72 Nobel laureates, between the top six universities in the world, 500 ha campus), il Campanile (Sather Tower (61 bells (full concert carillon) and clock tower). 1914, 94 m, 7 floors, observation deck on the 8th floor, inspired by il Campanile (850, 1514, 1912, 99 m) di San Marco (1084), Venezia (421, Venice), Italy (900 BC)).

Table of Contents

Preface 3
Table of Contents 5
3543 Dialog 1: Terraforming origins 9
3543 Dialog 2: Starting of Terraforming 11
3543 Dialog 3. Mars Terraforming 13
3543 Dialog 4. Improving temperature on Mars 16
3543 Dialog 5. Restoring Mars' magnetic field 18
   Back on 18
      In Region R0: from Paris (France) to N'Djamena (Chad) 20
      In Region R1: from Zagreb (Croatia) to Bujumbura (Burundi) 22
      In Region R2: from Kiev (Ukraine) to Baghdad (Iraq) 24
      In Region R3: from Riyadh (Saudi Arabia) to Malé (Maldives) 26
      In Region R4: from Bishkek (Kyrgyzstan) to Brahmapur (India) 28
      In Region R5: from Kathmandu (Nepal) to Dehong (China) 30
      In Region R6: from Bangkok (Thailand) to Chita (Russia) 32
      In Region R7: from Nanchang (China) to Melbourne (Australia) ... 34
      In Region R8: from Anchorage (Alaska, USA) to Lima (Peru) 36
      In Region R9: from La Paz (Bolivia) to London (United Kingdom) .. 38
   5.6. Five Assistants and terraforming 55
   5.7. The Honorific World Observer and terraforming 57
   5.8. Small World Government with 7 Departments 59

- Tax Department 60

- Treasury 62

- People Assistance Department 65

- Medical Department 67

- Police 70

- Education Department 73

- Science & Technology Department 75

3543 Dialog 6. Moon terraforming 78

3543 Dialog 7. Bio-robots for terraforming 84

3543 Dialog 8. Deep space exploration finds many places ready for terraforming 86

3543 Dialog 9. Multi-loop coupling mechanisms for terraforming 88

3543 Dialog 10. Meteors and terraforming 90

3543 Dialog 11. Places for terraforming 92

3543 Dialog 12. Ceres for terraforming 94

3543 Dialog 13. Vesta for terraforming 96

3543 Dialog 14. Pallas ready for terraforming 98

3543 Dialog 15. Hygiea can't wait terraforming 100

3543 Dialog 16. Interamnia is ready for terraforming 101

3543 Dialog 17. Europa wants fast terraforming 103

3543 Dialog 18. Sylvia graciously asks for terraforming 105

3543 Dialog 19. Ceramics for terraforming 107

3543 Dialog 20. Eunomia wants terraforming now 110

3543 Dialog 21. Euphrosyne likes terraforming 111

3543 Dialog 22. Cybele is on the list for terraforming 112

3543 Dialog 23. Juno is longing for terraforming 114

3543 Dialog 24. Patentia patiently waits for terraforming .. 116

3543 Dialog 25. Bamberga loves terraforming 118

3543 Dialog 26. Psyche is looking for terraforming 120

3543 Dialog 27. Thisbe – to be or not to be terraformatted 122

3543 Dialog 28. Silicones for terraforming 124

3543 Dialog 29. Doris is thinking terraforming 126

3543 Dialog 30. Fortuna needs assistance for terraforming 128

3543 Dialog 31. Themis is optimistic about terraforming 130

3543 Dialog 32. Aurora graciously waits for terraforming 131

3543 Dialog 33. Amphitrite smiles to terraforming 133

3543 Dialog 34. Egeria is eager for terraforming 135

3543 Dialog 35. Elektra wants electricity with terraforming............... 137

3543 Dialog 36. Iris is longing for terraforming 139

3543 Dialog 37. Hebe 140

3543 Dialog 38. Eugenia 142

3543 Dialog 39. Metis is patiently waiting for terraforming............... 145

3543 Dialog 40. Eleonora is euphoric about terraforming 147

Bibliography 149

UK, London: From the Bow Street, the northeast façade of the Royal Opera House at Covent Garden (1732, 1808, 1858, 1999, capacity 2,256). In 1734, Covent Garden presented its first ballet, Pygmalion. On 14 January 1947, the Covent Garden Opera Company gave its first performance of Carmen (1875, opera in four acts, based on a novella of the same title by Prosper Mérimée (1803-1870 (age 67))) by French composer Georges Bizet (1838-1875 (age 36)).

3543 Dialog 1: Terraforming origins

MARS: Tell me please, what are the origins of terraforming?

MOON: Well, as you know, people on Earth started long ago thinking about moving to the stars, starting with you and me, but this was not so easy. Around 1910, some writers and theorists began to consider the idea of transforming other space entities (planets, moons, asteroids, etc.) to be as analogous as possible to Terra.

MARS: But the people on Terra obviously needed a much better system on their planet, in order to be able to start working on terraforming.

MOON: Certainly, they had to implement a new system, or Constitution, which starts with this:

We, the People on this Earth, in order to
    1.1 - completely eliminate war and any type of conflicts,
    1.2 - have a peaceful and harmonious world,
    1.3 - have freedom, dignity, good families and respect,
    1.4 - have good health and good education,
    1.5 - have a friendly atmosphere and prosperity,
    1.6 – have the safety and wellbeing of all the people in the world as the highest priority,
    1.7 – use the best peaceful results, experience and knowledge of all current countries,

establish this Constitution of the World.

MARS: It is immediately noticeable that their first priority was the elimination of war and other types of conflicts, which they had for millennia.

MOON: Yes, indeed, and building on this they finally have a peaceful and harmonious world, freedom & order, good health, good education, good economy, dignity, good families, respect, and prosperity for all.

MARS: Do you remember what James Madison said?

MOON: Certainly: "The means of defense agst. foreign danger, have been always the instruments of tyranny at home."
"The advancement and diffusion of knowledge is the only guardian of true liberty."

MARS: Then, naturally, good health and good education are high priorities.

MOON: Nice to see that the safety and wellbeing of all the people on Earth is mentioned as the highest priority.

Italy, Roma - Arco di Costantino (312, left), and Amphitheatrum Flavium (Colosseum, 80 AD, right), from Via di San Gregorio.

3543 Dialog 2: Starting of Terraforming

MARS: I noticed that they started terraforming with you, my friend MOON.

MOON: Sure enough, being the closest to our beloved Earth, the specialists there are working hard to make me as comfortable as possible for people from Earth.

MARS: I understand that for better life on Earth it was necessary to use the best peaceful results, experience and knowledge, which was available long ago.

MOON: Certainly, now all the people on Earth are happy citizens of only one country, called Peaceful Terra, with total area of over 509 M km^2, and land area over 148 M km^2. No conflicting countries.

MARS: I remember Earth was full of all kinds of rules and regulations.

MOON: Not anymore – now all the rules – not more than 2,000, on maximum 1,000 pages - on Earth are established by the people and their elected Advisers.

MARS: Do you remember Churchill?

MOON: But of course: "If you have ten thousand regulations you destroy all respect for the law."

Paris (250 BC): l'Hôtel de Ville (City Hall since 1357, King Francis I started this building in 1533, finished 1628, 1873-1892.

3543 Dialog 3. Mars Terraforming

MARS: Of course that terraforming me is a very intense activity.

MOON: Indeed, one of the first priority is to increase your average temperature, which is around minus 62°C.

MARS: Sure, but now tell me what did they do on Peaceful Terra for easier administration?

MOON: For easier administration, Peaceful Terra was long ago only administratively divided in 10 simple and friendly regions, called R0, R1,…, R9, which are delimited by meridians (or line of longitudes).

Each region has a pair of capitals plus an outside city, for better and more homogenous management (all change every year). For example, the first implementation was:

R0 between meridians 0 and 15^0 E, capitals: Bern (Switzerland), Libreville (Gabon), and Oxford (UK).
R1: 15^0 E - 30^0 E, Warsaw (Poland), Pretoria (South Africa) and Miami (FL, USA).
R2: 30^0 E - 45^0 E, Moscow (Russia), Cairo (Egypt), and Grenoble (France).
R3: 45^0 E - 75^0 E, Astana (Kazakhstan), Karachi (Pakistan), and Montpellier (France).
R4: 75^0 E - 85^0 E, New Delhi (India), Novosibirsk (Russia), and Magdeburg (Germany).
R5: 85^0 E - 100^0 E, Krasnoyarsk (Russia), Urumqi (China), and Avignon (France).
R6: 100^0 E - 115^0 E, Jakarta (Indonesia), Beijing (China), and Neuchâtel (Switzerland).
R7: 115^0 E - 180^0, Tokyo (Japan), Sydney (Australia), and Malmö (Sweden).
R8: 180^0 - 70^0 Washington (USA), Mexico City (Mexico), and Bellinzona (Switzerland).
R9: 70^0 W – 0 Halifax (Canada), Brasilia (Brazil), and Biel (Switzerland).

USA, Washington (1790), National Gallery of Art (1937, National Mall)

Each of the 10 regions was divided by meridians in 10 sub-regions S00, , S99.

Each of the 100 sub-regions was divided in 10 districts D000, D001, , D999, each with about 7.7 M people, and each of the districts has their current small and big cities.

Having telework, many people have a northern residence and a southern residence, seasonally moving from one to the other, to avoid extreme cold or heat, and having the same hour.

All the oceans belong to the regions defined above, therefore are maintained by those regions, to be free of any piracy or other bad activity – World Police help when necessary.

UK, London: The northeast exterior of Paul Hamlyn Hall (old Covent Garden flower market), southeast of the Royal Opera House (right

3543 Dialog 4. Improving temperature on Mars

MARS: It is really nice to have a better temperature.

MOON: Yes, they release special dust particles into the atmosphere to initiate a greenhouse effect and block heat.

MARS: This is a big success! Now tell me how Peaceful Terra without borders is?.

MOON: You remember that for thousands of years people dreamed about not having borders – now, finally, this dream is reality.

There are just simple administrative delimitations, and all these delimitations between regions, as well as between sub-regions, are flexible – they are changed after each census (5 years), for maintaining a balanced number of people in all regions.

Because all the people are in the same country, it is normal to modify a little its regions, for better administration, to make everybody happy.

It is well understood that there were some difficulties in the beginning, like in all beginnings, but with calm, patience, perseverance and hard work, the things improved really fast, and all the people enjoy a better life.

MARS: Flexible delimitations look good.

MOON: Yes, to eliminate big differences between regions, it is really great to change these delimitations after each census (5 years). These changes be mostly in computers, and people benefit having better services from the government. For example, the driver license are valid everywhere. The same story for the car inspection.

UK, London: Inside the Covent Garden Market, with an opera soprano performing (center right down).

3543 Dialog 5. Restoring Mars' magnetic field

MARS: This magnetic field is a serious problem.

MOON: Yes, they are working on the core of Mars, and also on a man-made magnetic field placed between the Sun and Mars.

Back on Peaceful Terra, there are four levels of world management; at the local level, if needed, it could be one or two more levels of local managers (mayors, town managers, county managers – all levels of management are friendly, helpful, fast, polite, modest and smart):

1,000 L1 friendly managers for the 1,000 districts, who are supervising and assisting the mayors and town managers from their district. Each of the 1,000 L1 friendly managers are located in a central city from their districts – they could be the mayors of those cities, but with new responsibilities for the whole district.

MARS: Nice to see the emphasis on friendliness at the local level.

MOON: Yes, sometimes the local government were not as friendly as they should be, and now we can see a big change in better.

100 L2 friendly managers for the 100 sub-regions, who supervise and assist the 10 L1 managers of the 10 districts of each sub-region, for a total of about 77,000,000 people for each sub-region. These 100 L2 friendly managers move each month between the two capitals of each of the 100 sub-regions.

Italy, Venezia - The large yacht Vivaldi (left) tied up at the beautiful waterfront Riva dei Sette Martiri, near a bridge over a canal by the Fondamenta Rio della Tana, and Via Giuseppe Garibaldi is in the middle.

In the beginning these capitals were:

In Region R0: from Paris (France) to N'Djamena (Chad)

- The sub-region R00 had the capitals Paris (France) and Niamey (Niger) – assistance from Magdeburg (Germany).
- The sub-region R01 had the capitals Brussels (Belgium) and Porto-Novo (Benin) - assistance from Toronto (Canada).
- The sub-region R02 had the capitals Amsterdam (Netherlands) and Algiers (Algeria) - assistance from Graz (Austria).
- The sub-region R03 had the capitals Luxembourg (Luxembourg) and Sao Tome (Sao Tome and Principe) - assistance from Adelaide (Australia).
- The sub-region R04 had the capitals of Abuja (Nigeria) and Bochum (Germany) - assistance from Nikko (Japan).
- The sub-region R05 had the capitals Malabo (Equatorial Guinea), and Zürich (Switzerland) - assistance from Leeds (UK).
- The sub-region R06 had the capitals Oslo (Norway) and Tunis (Tunisia) - assistance from Sheffield (UK).
- The sub-region R07 had the capitals Roma (Italy) and Luanda (Angola) - assistance from Yamagata (Japan).
- The sub-region R08 had the capitals in Berlin (Germany) and Tripoli (Libya) - assistance from New York (USA).
- The sub-region R09 had the capitals Prague (Czech Republic) and N'Djamena (Chad) - assistance from Brisbane (Australia).

MARS: There are many great cities here.

MOON: And they work with smaller cities, to have a better understanding of different groups of people, for better assistance to all people of each sub-region.

Rome: Center: Columna Traiani (113 AD) with a band (180 m) of carved reliefs, which winds around the Trajan's Column, regarding Trajan's Dacian war campaigns (101-102 and 105-106 AD). After Trajan's death, his 6 m statue was on top until 1587. His ashes and later those of his wife Plotina were placed in the base of the column. Left: Santissimo Nome di Maria al Foro Traiano (1751, the Church of the Most Holy Name of Mary at the Trajan Forum).

In Region R1: from Zagreb (Croatia) to Bujumbura (Burundi)

- The sub-region R10 had the capitals in Zagreb (Croatia) and Brazzaville (Congo) - assistance from Nantes (France).
- The sub-region R11 had the capitals in Vienna (Austria), Windhoek (Namibia) - assistance from Bilbao (Spain).
- The sub-region R12 had the capitals in Stockholm (Sweden), Bangui (Central African Republic) - assistance from Florence (Italy).
- The sub-region R13 had the capitals in Budapest (Hungary), Rundu (Namibia) - assistance from Monaco (Monaco).
- The sub-region R14 had the capitals in Belgrade (Serbia), Kananga (Democratic Republic of Congo) - assistance from Liverpool (UK).
- The sub-region R15 had the capitals in Athens (Greece), Mongu (Zambia) - assistance from Los Angeles (CA, USA).
- The sub-region R16 had the capitals in Helsinki (Finland) and Kolwezi (Democratic Republic of the Congo) - assistance from Montreal (Canada).
- The sub-region R17 had the capitals in Bucharest (Romania) and Gaborone (Botswana) - assistance from Philadelphia (PA, USA).
- The sub-region R18 had the capitals in Minsk (Belarus) and Maseru (Lesotho) - assistance from Orleans (France).
- The sub-region R19 had the capitals in Chisinau (Republic of Moldova) and Bujumbura (Burundi) - assistance from Hamburg (Germany).

MARS: Let's see, the enjoyable sub-region R16 had the capitals in Helsinki (Finland) and?

MOON: Kolwezi (Democratic Republic of the Congo) – it was a typo in a previous edition of the Constitution, and we apologize. The good news is that these capitals frequently change, for the benefit of all people.

USA: The George Washington Bridge (1962, 1450 m, spanning Hudson River from New York City to Fort Lee, New Jersey, with routes 95 and 80, near Exit 73 for Fort Lee and route 67.

USA, New York: W 42nd Street, near 8th Avenue, with the Chrysler Building (1930, 320 m, 77 floors, center-right far back).

In Region R2: from Kiev (Ukraine) to Baghdad (Iraq)

- The sub-region R20 had the capitals in Kiev (Ukraine) and Kigali (Rwanda) - assistance from Ottawa (Canada).
- The sub-region R21 had the capitals in Ankara (Turkey) and Khartoum (Sudan) - assistance from Salzburg (Austria).
- The sub-region R22 had the capitals in Lilongwe (Malawi) and Nicosia (Cyprus) - assistance from Dallas (TX, USA).
- The sub-region R23 had the capitals in Jerusalem (Israel) and Dodoma (Tanzania) - assistance from Strasbourg (France).
- The sub-region R24 had the capitals in Damascus (Syria) and Nairobi (Kenya) - assistance from Stuttgart (Germany).
- The sub-region R25 had the capitals in Krasnodar (Russia) and Addis Ababa (Ethiopia) - assistance from Marseille (France).
- The sub-region R26 had the capitals in Rostov-on-Don (Russia) and Asmara (Eritrea) - assistance from Leipzig (Germany).
- The sub-region R27 had the capitals in Stavropol (Russia) and Djibouti (Djibouti) - assistance from Zürich (Switzerland).
- The sub-region R28 had the capitals in Mosul (Iraq) and Moroni (Comoros) - assistance from Linz (Austria).
- The sub-region R29 had the capitals in Yerevan (Armenia) and Baghdad (Iraq) - assistance from Göttingen (Germany).

MARS: Nice to see Göttingen (Germany) helping Yerevan (Armenia) and Baghdad (Iraq), in the remarkable sub-region R29.

MOON: This is the beauty – people from different parts of the world help each other to get better together.

USA, New York: W 42nd Street, near 8th Avenue, with the Chrysler Building (1930, 320 m, 77 floors, center-right far back).

UK, London: On James Street at Long Acre, City of Westminster, Covent Garden Station, 100 m west of the Royal Opera House.

In Region R3: from Riyadh (Saudi Arabia) to Malé (Maldives)

- The sub-region R30 had the capitals in Riyadh (Saudi Arabia) and Mogadishu (Somalia) - assistance from Bonn (Germany).
- The sub-region R31 had the capitals in Baku (Azerbaijan) and Antananarivo (Madagascar) - assistance from Le Mans (France).
- The sub-region R32 had the capitals in Oral (Kazakhstan) and Tehran (Iran) - assistance from Pisa (Italy).
- The sub-region R33 had the capitals in Ashgabat (Turkmenistan) and Abu Dhabi (United Arab Emirates) - assistance from Wolfsburg (Germany).
- The sub-region R34 had the capitals in Magnitogorsk (Russia) and Muscat (Oman) - assistance from Toulouse (France).
- The sub-region R35 had the capitals in Chelyabinsk (Russia) and Herat (Afghanistan) - assistance from Basel (Switzerland).
- The sub-region R36 had the capitals in Tyumen (Russia) and Kandahar (Afghanistan) - assistance from Nagoya (Japan).
- The sub-region R37 had the capitals in Dushanbe (Tajikistan) and Labytnangi (Russia) - assistance from Limoges (France).
- The sub-region R38 had the capitals in Tashkent (Uzbekistan) and Kabul (Afghanistan) - assistance from Rostock (Germany).
- The sub-region R39 had the capitals in Islamabad (Pakistan) and Malé (Maldives) - assistance from La Rochelle (France).

MARS: Noteworthy - Pisa (Italy) was working with Oral (Kazakhstan) and Tehran (Iran) in the stunning sub-region R32.

MOON: Bringing together different civilizations was very productive for all people.

MARS: The sub-region R38 had the capitals in Kabul (Afghanistan) and?

MOON: Tashkent (Uzbekistan). Nice to see Rostock (Germany) working together with Tashkent (Uzbekistan) and Kabul (Afghanistan) – a really nice combination of cities.

Japan, 13 km north-east from Mount Fuji, the easternmost and largest of the five lakes, Lake Yamanaka is also the third highest lake in Japan, standing at 980 meters above sea level.

In Region R4: from Bishkek (Kyrgyzstan) to Brahmapur (India)

- The sub-region R40 had the capitals in Bishkek (Kyrgyzstan) and Jaipur (India) - assistance from Osaka (Japan).
- The sub-region R41 had the capitals in Akola (India) and Kashgar (China) - assistance from Genoa (Italy).
- The sub-region R42 had the capitals in Almaty (Kazakhstan) and Coimbatore (India) - assistance from Perth (Australia).
- The sub-region R43 had the capitals in Kuybyshev (Russia) and Agra (India) - assistance from Fukuoka (Japan).
- The sub-region R44 had the capitals in Vertikos (Russia) and Nagpur (India) - assistance from Coral Bay (Australia).
- The sub-region R45 had the capitals in Chennai (India) and Colombo (Sri Lanka) - assistance from Sapporo (Japan).
- The sub-region R46 had the capitals in Lucknow (India) and Fedosikha (Russia) - assistance from Niigata (Japan).
- The sub-region R47 had the capitals in Bilaspur (India) and Kolpashevo (Russia) - assistance from Albany (Australia).
- The sub-region R48 had the capitals in Visakhapatnam (India) and Barnaul (Russia) - assistance from Hiroshima (Japan).
- The sub-region R49 had the capitals in Brahmapur (India) and Tomsk (Russia) - assistance from Yokohama (Japan).

MARS: What a nice combination - Hiroshima (Japan) working with Visakhapatnam (India) and Barnaul (Russia), in the attractive sub-region R48.

MOON: Yes, people with different experiences working together for a better future for all.
Churchill said it right: If we open a quarrel between past and present, we shall find that we have lost the future.

UK, London: From the British Museum (1753), looking northwest to a nice building after Great Russell Street.

USA, New York: On W 42nd St, the northeast façade of the New York Public Library (1902).

In Region R5: from Kathmandu (Nepal) to Dehong (China)

- The sub-region R50 had the capitals in Kathmandu (Nepal) and Patna (India) - assistance from Kobe (Japan).
- The sub-region R51 had the capitals in Bayingol (China) and Novokuznetsk (Russia) - assistance from Vichy (France).
- The sub-region R52 had the capitals in Thimphu (Bhutan) and Dhaka (Bangladesh) - assistance from Jena (Germany).
- The sub-region R53 had the capitals in Lhasa (China) and Achinsk (Russia) - assistance from Reims (France).
- The sub-region R54 had the capitals in Abakan (Russia) and Kumul (China) - assistance from Fribourg (Switzerland).
- The sub-region R55 had the capitals in Kyzyl (Russia) and Dibrugarh (India) - assistance from Denmark (Australia).
- The sub-region R56 had the capitals in Bassein (Myanmar) and Tinsukia (India) - assistance from Chiba (Japan).
- The sub-region R57 had the capitals in Yushu City (China) and Tinskoy (Russia) - assistance from Klagenfurt (Austria).
- The sub-region R58 had the capitals in Jiuquan (China) and Medan (Indonesia) - assistance from Lucerne (Switzerland).
- The sub-region R59 had the capitals in Chiang Mai (Thailand) and Dehong (China) - assistance from Mulhouse (France).

MARS: Look at this - Lucerne (Switzerland) working with Jiuquan (China) and Medan (Indonesia) in the astonishing sub-region R58.

MOON: It is impressive indeed – they certainly produce great results for everybody.

UK, London: At 31 Endell Street, Covent Garden, The Cross Keys pub, 300 m northwest from the Royal Opera House.

USA, New York: On W 42nd St at Avenue of the Americas, looking northwest at the Bryant Park (left), Grace building (right), Bank of America (next).

In Region R6: from Bangkok (Thailand) to Chita (Russia)

- The sub-region R60 had the capitals in Bangkok (Thailand) and Kuala Lumpur (Malaysia) - assistance from Besançon (France).
- The sub-region R61 had the capitals in Vientiane (Laos) and Singapore – assistance from Freiburg im Breisgau (Germany).
- The sub-region R62 had the capitals in Phnom Penh (Cambodia) and Irkutsk (Russia) – assistance from Baden (Switzerland).
- The sub-region R63 had the capitals in Palembang (Indonesia), Hanoi (Vietnam) – assistance from Thun (Switzerland).
- The sub-region R64 had the capitals in Ulan Bator (Mongolia) and Ulan-Ude (Russia) – assistance from Chaumont (France).
- The sub-region R65 had the capitals in Cirebon (Indonesia) and Nanning (China) – assistance from Vaduz (Lichtenstein).
- The sub-region R66 had the capitals in Pontianak (Indonesia) and Baotou (China) – assistance from Lugano (Switzerland).
- The sub-region R67 had the capitals in Surakarta (Indonesia) and Yichang (China) – assistance from Thonon-les-Bain (France).
- The sub-region R68 had the capitals in Surabaya (Indonesia) and Changsha (China) – assistance from Burgdorf (Switzerland).
- The sub-region R69 had the capitals in Chita (Russia) and Hong Kong (China) – assistance from Colmar (France).

MARS: Look at Besançon (France) working with Bangkok (Thailand) and Kuala Lumpur (Malaysia), in the lovely sub-region R60.

MOON: Yes, I'm glad to see them together, improving everybody's life.

MARS: Vita non est vivere, sed valere vita est.

MOON: Life is not being alive, but being well.

From a bus on Oxford Street at South Molton St (right), looking east to Tissot store, and many other stores.

USA, New York: At 401 Fifth Ave at E 37th St, looking south, TD Bank in a classic nice building.

In Region R7: from Nanchang (China) to Melbourne (Australia)

- The sub-region R70 had the capitals in Bandar Seri Begawan (Brunei Darussalam) and Nanchang (China) – assistance from Turku (Finland).
- The sub-region R71 had the capitals in Krasnokamensk (Russia) and Jinan (China) – assistance from St. Gallen (Switzerland).
- The sub-region R72 had the capitals in Baguio City (Philippines) and Hangzhou (China) – assistance from Dole (France).
- The sub-region R73 had the capitals in Manila (Philippines) and Taipei (Taiwan, China) – assistance from Metz (France).
- The sub-region R74 had the capitals in Kupang (Indonesia) and Shanghai (China) – assistance from Davos (Switzerland).
- The sub-region R75 had the capitals in Pyongyang (North Korea) and Seoul (South Korea) – assistance from Versailles (France).
- The sub-region R76 had the capitals in Vladivostok (Russia) and Busan (South Korea) – assistance from Innsbruck (Austria).
- The sub-region R77 had the capitals in Kyoto (Japan) and Khabarovsk (Russia) – assistance from Germering (Germany).
- The sub-region R78 had the capitals in Nagoya (Japan) and Komsomolsk-on-Amur (Russia) – assistance from Venice (Italy).
- The sub-region R79 had the capitals in Sendai (Japan) and Melbourne (Australia) – assistance from St. Moritz (Switzerland).

MARS: One cannot have a better combination than this - Versailles (France) working with Pyongyang (North Korea) and Seoul (South Korea) in the marvelous sub-region R75, for the benefit of all people.

MOON: Yes, it be a great success.
Confucius helps us: Success depends upon previous preparation.

Finland, Helsinki Central railway station (1907 – 1914), on Brunnsgatan, in the city center.

UK, London: Wellington (1769-1852) Arch (1830, four-horse chariot 1912), at southeast corner of Hyde Park, and western corner of Green Park

In Region R8: from Anchorage (Alaska, USA) to Lima (Peru)

- The sub-region R80 had the capitals in Uelen (Russia) and Anchorage (Alaska, USA), – assistance from Zug (Switzerland).
- The sub-region R81 had the capitals in Vancouver (Canada) and San Jose (CA, USA) – assistance from Odense (Denmark).
- The sub-region R82 had the capitals in Vernon (Canada) and Los Angeles (CA, USA) – assistance from Amstetten (Austria).
- The sub-region R83 had the capitals in Calgary (Canada) and Tijuana (Mexico) – assistance from Chur (Switzerland).
- The sub-region R84 had the capitals in Hermosillo (Mexico) and Tucson (AR, USA) – assistance from Bergen (Norway).
- The sub-region R85 had the capitals in Chihuahua (Mexico) and Regina (Canada) – assistance from Gothenburg (Sweden).
- The sub-region R86 had the capitals in San Luis Potosi City (Mexico) and Winnipeg (Canada) – assistance from Yverdon-les-Bains (Switzerland).
- The sub-region R87 had the capitals in Tulsa (OK, USA) and Veracruz (Mexico) – assistance from Bregenz (Austria).
- The sub-region R88 had the capitals in Memphis (TN, USA) and San José (Costa Rica) – assistance from Uppsala (Sweden).
- The sub-region R89 had the capitals in Lima (Peru) and Boston (MA, USA) – assistance from Tampere (Finland).

MARS: Now another wonderful combination - Zug (Switzerland) working with Uelen (Russia) and Anchorage (Alaska, USA), in the attractive sub-region R80.

MOON: One can expect really astonishing results, good for all people.
Here it is good to remember Jefferson: "All tyranny needs to gain a foothold is for people of good conscience to remain silent."

USA, Boston (founded in 1630): tall ships from many countries, at the Boston Fish Pier (opened in 1915).

Rome. North-west view of Rome from Altare della Patria (1925), with Via del Teatro di Marcello (13 AD, down left), and Basilica Papale di San Pietro in Vaticano (1506, center back).

In Region R9: from La Paz (Bolivia) to London (United Kingdom)

- The sub-region R90 had the capitals in La Paz (Bolivia) and Bangor (Maine, USA) – assistance from Aosta (Italy).
- The sub-region R91 had the capitals in Caracas (Venezuela) and Road Town (British Virgin Islands) – assistance from Obergoms (Switzerland).
- The sub-region R92 had the capitals in Buenos Aires (Argentina) and Fort-de-France (Martinique) – assistance from Freudenstadt (Germany).
- The sub-region R93 had the capitals in Amarscion (Paraguay) and Montevideo (Uruguay) – assistance from Winterthur (Switzerland).
- The sub-region R94 had the capitals in Cayenne (French Guiana), St. John's (Canada) – assistance from Novara (Italy).
- The sub-region R95 had the capitals in Rio de Janeiro (Brazil) and Dakar (Senegal) – assistance from Toyama (Japan).
- The sub-region R96 had the capitals in Freetown (Sierra Leone) and Lisbon (Portugal) – assistance from Kawasaki (Japan).
- The sub-region R97 had the capitals in Bamako (Mali) and Athlone (Ireland) – assistance from Ulm (Germany).
- The sub-region R98 had the capitals in Yamoussoukro (Cote d'Ivoire) and Madrid (Spain) – assistance from Okayama (Japan).
- The sub-region R99 had the capitals in Ouagadougou (Burkina Faso) and London (United Kingdom) - assistance from Vaasa (Finland).

MARS: Breath taking - Toyama (Japan) working with Rio de Janeiro (Brazil) and Dakar (Senegal), in the stupendous sub-region R95.

MOON: I cannot find words – they certainly significantly contribute to the success of this Constitution of the World! Between Toyama and Rio de Janeiro there are over 18,000 km, and between Rio de Janeiro (Brazil) and Dakar (Senegal), over 5,000 km, and with video and phone contacts, the collaboration work fine.

Finland, Helsinki: a Baltic Sea canal from west to east, near Ruoholahdenpuisto, seen from a bridge on Bottenhavsgatan, near Helsinki Conservatory of Music (left).

UK, London: From a bus going southwest on Pall Mall, Royal Opera Arcade Gallery (center, hosting art exhibitions and events).

Ten L3 friendly managers for the 10 regions, who supervise and assist the 10 L2 managers of the 10 sub-regions of each region.

MARS: These 10 regions are really big.

MOON: Yes, they require effective and energetic management.

- The Region R0 had the first capitals in Bern (Switzerland) and Libreville (Gabon) – assistance from Oxford (UK). For better quality and consistency of the management, we'll have the first two cities from the region R0, and the third city from outside. Actually, being inside the same country Terra, any city, sub-region or region can ask for advice or help from anybody.

- The Region R1 had the first capitals in Warsaw (Poland) and Pretoria (South Africa) – assistance from Miami (FL, USA).

- The Region R2 had the first capitals in Moscow (Russia) and Cairo (Egypt) – assistance from Grenoble (France).

- The Region R3 had the first capitals in Astana (Kazakhstan) and Karachi (Pakistan), – assistance from Montpellier (France).

- The Region R4 had the first capitals in New Delhi (India) and Novosibirsk (Russia) – assistance from Magdeburg (Germany).

- The Region R5 had the first capitals in Krasnoyarsk (Russia) and Urumqi (China) – assistance from Avignon (France).

- The Region R6 had the first capitals in Jakarta (Indonesia) and Beijing (China) – assistance from Neuchâtel (Switzerland).

- The Region R7 had the first capitals in Tokyo (Japan) and Sydney (Australia) – assistance from Malmö (Sweden).

- The Region R8 had the first capitals in Washington (USA) and Mexico City (Mexico) – assistance from Bellinzona (Switzerland).

- The Region R9 had the first capitals in Halifax (Canada) and Brasilia (Brazil) – assistance from Biel (Switzerland).

Japan, Tokyo, Shinjuku: suspended streets between tall buildings.

MARS: Really global - Malmö (Sweden) working with Tokyo (Japan) and Sydney (Australia) in the inspiring region R7.

MOON: They enhance the beauty of the world. Malmö (Sweden) to Tokyo (Japan) is over 8,500 km, and Tokyo (Japan) to Sydney (Australia) is over 7,700 km.

USA, Boston (founded in 1630): on a visiting tall ship, at the Boston Fish Pier (opened in 1915).

L4 very friendly 10 Advisers of the world, who supervise and assist the 10 L3 managers of the 10 regions of the Earth.

MARS: Now we are at the top level – great responsibility.

MOON: Indeed, the 10 Advisers calmly and intelligently manage the world, for the benefit of the people on the planet.
Here Goethe helps: All intelligent thoughts have already been thought; what is necessary is only to try to think them again.

Germany - 23 March 1978, Freibourg im Breisgau (1120 by Duke Berthold III of Zähringen (1085-1122), elevation 278 m, the south façade of Freiburger Münster (cathedral, 1200, 116 m, J. S. Bach (1685-1750) performed here).

The L4 very friendly 10 Advisers of the world are located each in one the ten Regions R0, R1,…, R9. For example, in the beginning, for the first month (then changing every month), the ten Advisers of the world were located in:

- in R0: Barcelona (Spain)
- in R1: Benghazi (Libya)
- in R2: Addis Ababa (Ethiopia)
- in R3: Hyderabad (Pakistan)
- in R4: Bhopal (India)
- in R5: Mandalay (Myanmar)
- in R6: Nanchong (China)
- in R7: Khabarovsk (Russia)
- in R8: Houston (USA)
- in R9: Recife (Brazil)

MARS: Are they moving around?

MOON: Yes, and they change the location every month, to be close to as many people as possible – this mobility help them understand what the people want, and how to achieve the people's objectives fast and efficient.

These ten L4 Advisers are in permanent contact with each other, and with the L3 Advisers, for the best management of the world.

The ten L4 Advisers work by consensus only.

MARS: Tell me about a very important management rule.

MOON: Yes, the 10 Advisors must be talented enough to be able to negotiate fast any disagreements between them, and quickly arrive at the best common decision for the benefit of all people.

The ten L4 Advisers are elected from the 10 regions, and each of them be the First Adviser (***First among equals*** – from Latin: Primus inter pares) for one month, by rotation.

The First Adviser only coordinates the work of the other 9 Advisors for one month.

MARS: No question that these management rules create the necessary harmony for the world.

MOON: Yes, the 10 Advisers represent the world, they were elected to improve the world, therefore they must harmoniously work together for the benefit of all people.

UK, London: On Pall Mall East SW, the Wellington building (center), Strand Palace Hotel (center left), building on right built in 1834.

The ten L4 Advisers move each month from a first capital of a region to the second capital of another region, at random (or based on urgency, if an emergency occurred). This mobility is essential for having a long period of tranquility and harmony.

MARS: Very important to have good mobility.

MOON: Yes, the 10 Advisers were elected from their regions, but they must stay in all the regions, to understand the local problems of all the regions, in order to make decisions which benefit all the people.
The First Adviser, on the last day of each month, presents in writing for the world (no more than 5 standard pages) a clear and precise Monthly World Report, with a list of finished and unfinished tasks.

MARS: This is essential.

MOON: Indeed, a clear and precise Monthly World Report (5 pages), with a list of finished and unfinished tasks, help everybody understand what was done, what remains to be done, and what ideas are about doing them fast and correct.

The other 9 Advisers add their comments to the Monthly World Report (no more than half a page each - total report less than 9.5 pages).

MARS: Comments are always welcome.

MOON: Harmonious comments, which want to help, give new ideas, etc.

Finland, Helsinki: a government building on Abrahaminkatu, northeast of a large square

The top 10 Advisers manage Police and all other Departments.

MARS: Important to put Police first.

MOON: Yes, Police is the only Department which can use some force, in extreme situations, therefore there is need to have strict control on how to use this force, and actually to try to minimize the use of force, and maximize the use of calm dialog, medical assistance, and friendly approach.

For obvious uncooperative or improper attitude of one top Advisor X, the other 9 can replace X with X's number 2, and X receive appropriate medical treatment.

When vacancies happen for Advisors, the number 2 for those Advisors fill the vacancies.

All the activities of all Advisors are recorded in computers and videos, and on paper, for people to be able to see what they are doing.

Advisors at all levels work 40 hours/week, with 4 weeks of vacation, but many services (medical, police (firemen are part of the police), emergency, volunteers) are non-stop.

Advisors' compensation is the world annual average salary plus 4% of that world average salary, for level 4, + 3 % for level 3, and so on. They all are working hard to increase the world average salary, in order to get themselves an increase.

All the other world government employees have a compensation close to the average compensation of the people in the area where they are located.

All Advisors are free to speak about their administrative work, with modesty.

MARS: This modesty is really important.

MOON: Sure, all Advisors work for people, and only the people have a say about Advisers' administrative work.

MARS: These spending proposals are always problematic.

MOON: For this reason, they must be approved by all assistants (doctors, mathematicians, CEOs, engineers and teachers), and must have an already existing funding in the budget.

Rome: Center: Columna Traiani (113 AD) with a band (180 m) of carved reliefs. Left: Altare della Patria (1925).

Advisors (and all the others) cannot declare war, reprisals or capture land or water.

Advisors (and all the others) cannot raise and support armies, navy, or any military forces.

MARS: These statements are really fundamental.

MOON: Indeed, all the people want peace, freedom and prosperity, therefore it is natural to have these very clear, simple and fundamental requirements.

And we remember Hoover: Older men declare war. But it is youth that must fight and die.

At least 7 of the top 10 Advisers should be present every working day.

MARS: Normal requirements.

MOON: Yes, the government is, simple and modest.

Washington (1790), National Archives and Records Administration building (1935), on Constitution Avenue.

In order to better know the world government, to help it, and, especially, to improve it, all able people of the world are working as volunteers at least one day per year in each of the seven departments.

MARS: Excellent idea!

MOON: Indeed, this brings the government close to the people, which pay all their salaries and expenses – a close cooperation between people and their friendly government is a necessity for a harmonious world.
Good to remember Anatole France: "If a million people say a foolish thing, it is still a foolish thing."
After each Monthly World Report, a public opinion survey about the report should be taken, and presented to all Advisors.

MARS: Clearly useful.

MOON: Sure, people must express their opinions on the work of their Advisors, and Advisors must keep in mind that they have to work hard and smart to fulfill their obligations for all people.
Cicero comes handily: "Nature has planted in our minds an insatiable longing to see the truth."

Japan, 13 km north-east from Mount Fuji, the easternmost and largest of the five lakes, Lake Yamanaka is also the third highest lake in Japan, standing at 980 meters above sea level.

France, Paris: Jardin du Carrousel and Pavillion de Marsan, at the west end of the north part of Musée du Louvre; it also was located at the northern end of Palais des Tuileries (1564 – 1883 demolished).

All activities of the Advisors, and others from the small World Government, are available to the people on our excellent websites.

The top 10 Advisers (and all the others) collaborate via e-mail, telephone, videoconferences, mail, or face to face, when needed, to produce practical results for all people, very fast.

MARS: Bottom line - produce practical results for all people, very fast.

MOON: This is it – no words, just practical results.

MARS: Facta non verba.

MOON: Deeds, not words.

Italy, Venezia - On the west façade of Palazzo Ducale, above the Porta della Carta, we see the Winged Lion and on top raises a statue of Justice.

USA, New York: On Fifth Avenue at E 40th St, looking southwest at Mid-Manhattan Library, a New York Public Library (1895, 1908, 87 branches (Carnegie libraries (Andrew Carnegie (1835-1919))), 53 millions of books and other items, the 2nd largest public library in the United States (behind the Library of Congress), and the fourth largest in the world (after British Library (170 M), Library of Congress (160 M), and Library and Archives Canada (54 M)) image archive (left), having thousands of photos, posters, illustrations, and other images.

5.6. Five Assistants and terraforming

Each Advisor, and each manager at all levels, has 5 immediate assistants:
1) a mathematician for finance and all other calculations,
2) a medical doctor for keeping everybody healthy, calm, polite, friendly and optimist,
3) a CEO for good management,
4) an engineer for all practical projects, and
5) a teacher for education, training and related areas.

MARS: Well, this is the real hope.

MOON: Certainly, only by having highly knowledgeable and talented people, it is possible to implement the many complex and pressing tasks, which are at the door of the World Government.

MARS: Dum spiro, spero.

MOON: While I breathe, I hope.

The five assistants play a key role, because they are highly qualified professionals, who actually carry on the practical management of the world.

The five assistants' integrity, professionalism and friendliness significantly improve the quality of the world and local governments.

The five assistants are really the experts. They assist the Advisors and all levels of management, in order to have an efficient, correct and professional working of the world government at all levels, including for terraforming.

MARS: Clear statements.

MOON: Yes, they give real hope that the world government is working how the people expect.

MARS: Aut disce aut discede.

MOON: Either learn or leave.

USA, New York: On Fifth Ave at East 34th St, the Graduate Center of the City Univ. of New York (left), a small building across Empire State Building.

5.7. The Honorific World Observer and terraforming

A Honorific World Observer is quietly elected by direct vote, for only one 3 years term, with the main duty to observe that the top 10 Advisers efficiently perform their duties, and keep their words, inclusive regarding terraforming – if they don't, they will be changed.

MARS: Now comes the Honorific World Observer.

MOON: Yes, this Observer is needed to monitor and assist the 10 Advisors – they all must efficiently perform their duties, and keep their words.
For managers and for everybody else, keeping their word is a serious and strict requirement.
The Honorific World Observer has this responsibility for the top 10 Advisors, but all people pay attention to this. Words are now important and respected.

MARS: Nice to see that the words count again.

MOON: Sure, it is strict necessary to have the words again important and respected.

Finland, Helsinki: beautiful buildings on Mikonkatu, close to Aleksanterinkatu, 200 m south-east of the Helsinki Central Railway Station.

5.8. Small World Government with 7 Departments

All the employees of the World Government are temporary, and must reapply for their positions every year.
There is no need for unions.
The World Government be limited to:
1) the Office of the Honorific Observer (less than 10 employees),
2) the Office of the top ten Advisors (less than 100 employees), and
3) 7 small departments.

MARS: Clear and useful statements.

MOON: Yes, the World Government works for all people, and should have clear limitations, not to waste people's money and time.
The World Government had these 7 small departments:

MARS: 7 is a good number.

MOON: Yes, not too small, and, especially, not too big.

- Tax Department

- Collects taxes of 15% of the income of people and revenue of companies.
- The Manager of the Tax Department is appointed for a three-year term by the World 10 Advisers.
- The number of employees must be under 50,000, with excellent computers, and advanced software.

MARS: Good beginning.

MOON: Yes, the Tax Department is important, brings revenue to the World Government, and the government must use this money for people's benefit.

Rome. Isola Tiberina (left), Pons Cestius (27 BC, right up), viewed from Ponte Garibaldi, 1888, joining Regola & Trastevere (right).

Rome. Down: a part of Forum Augustum (2 BC). Back: a part of Forum Traiani (113 AD), including Columna Traiani (113, center back), with a band (180 m) of carved reliefs, which winds around the Trajan's Column, describing Trajan's Dacian war campaigns (101-102 and 105-106 AD). After Trajan's death, his 6 m statue was on top until 1587. His ashes and later those of his wife Plotina were placed in the base of the column.

- Treasury

Treasury control all the financial issues, including:
- antitrust
- fiscal service
- financial cooperation
- financing bank
- world reserve system
- world budget using only revenue, no borrowing, and spending only on strict necessary needs
– all the budgets, at all levels, have a 2% surplus, which be returned to the taxpayers
- register of all government papers and activities
- archives and records
- assist all people to have savings accounts for old age (the old age be starting around 70), and 10% of their income should automatically go to their savings accounts. For those unable to work, their doctors and mathematicians decide case by case.
- bankruptcies, in general, are discouraged, and when strict necessary, be analyzed and solved, case by case, by the doctors, mathematicians and CEOs who worked with the people who asked the bankruptcy.
- encourage all families to assist their parents, grandparents, and great-grandparents.
- housing finance
- housing for all people
- no homelessness
- consumer financial protection
- pensions
- privacy
- personal savings
- personnel management
- general services for the world government
- each the 10 regions receive 2.5% of the world taxes - at least 30% of the money are sent to villages and cities.
- each of the 100 sub-regions receive 0.25% of the world taxes. At least 40% of the money is sent to villages and cities.

- The World Central Bank include all current central banks.
- The Special Credit Card (SCC) is issued by the World Central Bank.
- Advisors created a new world currency, named "coin", and all the other currencies were exchanged for coins. The World Central Bank implemented the details.
- The counterfeiting and all other bad things, which some sick people do, are medically treated (in specialized medical institutions when necessary), and those who did bad things pay all the expenses, and reimburse the victims. Victims always are very protected and helped to recover the losses from the attackers.

MARS: There are many relevant responsibilities in this Treasury Department.

MOON: Yes, like:
- world reserve system
- world budget using only revenue, no borrowing, and spending only on strict necessary needs
– all the budgets, at all levels, have a 2% surplus, which is returned to the taxpayers
- assist all people to have savings accounts for old age
- housing for all people

UK, London: On Broad Ct looking northeast, off Bow Street to the northeast, 50 m north of the Royal Opera House at Covent Garden (1732, 1808, 1858, 1999, capacity 2,256; in 1734, Covent Garden presented its first ballet, Pygmalion), the bronze statue Young Dancer, by the Italian-born (in Mestre, near Venice, in 1921) British sculptor Enzo Plazzotta (1921-1981 (age 60)). To the right up, five red telephone boxes, at 5 Broad Ct, a tourist attraction.

- People Assistance Department

It is assisting people in general, including:
- parent assistance
- dispute resolution
- in very simple disputes or culpa levis (ordinary negligence, like late payments, etc.), one single assistant decides within minutes, and all people go back to work
- census every 5 years
- election assistance every 20 months
- special credit cards
- people protection against abuses from anybody
- completely eliminated corruption, organized crime and drug trafficking
- all people in the world remained in their places, and the improvements have come to them. Those who want to move to other places, need first a special invitation from at least 10 people (not family related) where they want to move.
- all the Tribunals and related areas were transformed in people assistance services, based on friendliness, collaboration and goodwill.
- It is well understood that no excessive bail is required, no excessive fines imposed, no cruel and unusual punishments applied, but, at the same time, it is well understood that a person who did a bad thing receives the necessary corrective medical treatment, and reimburses all people who suffered damages, and the medical treatment. The victims always receive special attention.
- Nobility (King, Prince, etc.) continue to exist in some places, but they are not interfering with activities of the Advisors, and actually help them.
- food safety
- trash & recycling
- free commerce
- jobs assistance
- postal service
- labor safety and harmonious relations
- land, water

- volunteers
- fitness, sport, tourism, terraforming
- 10 world holidays: the normal 4 Earth events (2 solstices (around 21 June, around 21 December), and 2 equinoxes (around 21 March, around 21 September), Mother's Day on 1st May, Father's Day on 6 August, Children's Day on 6 November, Grandparents' Day on 6 February, and 2 optional days (like Thanksgiving or a Religious Day (Christmas), and New Year).

MARS: This new People Assistance Department is a real joy!

MOON: Yes, it has plenty of very useful responsibilities:

MARS: Cicero helps here:

MOON: Never go to excess, but let moderation be your guide. Faithfulness and truth are the most sacred excellences and endowments of the human mind.

- Medical Department

It manages all medical and healthcare related areas, including:
- human services
- conflict resolution
- families, children, elderly
- medicine approval
- disease control and prevention
- medical doctors and assistants make regular home visits, at least once a year, to all people, to keep them healthy, and to prevent illnesses.
- medical research: cancer, heart, lung, blood, arthritis, surgical robotics, connected computers for healthcare, etc.
- healthy homes, streets, stores, working places, terraforming, etc.
- healthy aging
- all misunderstandings, disagreements or conflicts of any nature are treated by medical personnel (with police help when strict necessary), until all is back to normal.
- no prisons are necessary, only specialized medical institutions (in simple cases, the places where the treated people live can be used, with the necessary limitations and surveillance)
- If a person X is considered that did a bad thing, X has, within 3 days, a discussion with one or more doctors and other assistants, and is informed of the nature and cause of the bad thing, including witnesses against and for him. Then a decision is taken within other 3 days, by a group of doctors and other assistants. Victims of bad people always have priority to discuss their problems with one or more doctors and other assistants, and quick decisions are taken within 3 days, by a group of doctors and other assistants. Protection of victims has always priority.
- in order to better know the world government, to help it, and, especially, to improve it, all able people of the world work as volunteers at least one day per year in the local facility of this department, which has a special office for managing this volunteer work.

– all people have government medical insurance, and they can also have private medical insurance

– there are doctors working for the government 100%, or only part-time, or having only private practice, all with reasonable salaries and fees.

– there are government pharmaceutical institutions and private pharmaceutical companies, offering reasonable priced medicines, without advertising to the general public.

MARS: The very essential Medical Department is certainly full of responsibilities.

MOON: Of course.

MARS: Terence is here…

MOON: I am a man, and whatever concerns humanity is of interest to me.

Italy, Venezia - The left door on the west façade of Basilica Cattedrale Patriarcale di San Marco. Above the door we can see the Winged Lion, the symbol of St. Mark and of Venice, which holds the book quoting *"Pax Tibi Marce Evangelista Meus"* (Peace to you, Mark, my evangelist).

- Police

Police provide assistance for:
- accidents
- disasters
- complete elimination of nuclear, chemical and biological arms, firearms and explosives
- world complete security
- world cooperation
- conflict reduction and resolution
- investigations
- emergency assistance
- training
- delinquency prevention in general, and especially juvenile
- protection of Advisors, important government buildings, etc.
- extended surveillance and reconnaissance to prevent bad events
- fire protection
- volunteers to help police
- police be present at public meetings, services, shows, etc., in order to protect the public
- public order
- ensuring traffic safety
- completely eliminate corruption, organized crime and drug trafficking
- movement of people based on civilized rules
- assist and protect those who have encountered violence
- World Police and specialists from the former United Nations and Interpol are ready and very mobile for urgent and special operations, when they are needed.
- Police are the only department which have some small arms, in order to stop some very bad people (who are very sick).
- a small manufacturing and maintenance of arms unit is part of the Police Department, under strict control.
- Police work with medical personnel, mathematicians, CEOs, engineers, teachers and others, to make sure that all the people on the Planet are in good mental health, in order to prevent bad situations. This is also a major responsibility of all Advisors.

- prevention of bad events
- The Advisors allocate the necessary budget for Police, and Police assist people in need.

MARS: No question that police are important.

MOON: Certainly, and they become friends of all people, and help them:

Italy, Venezia - Two of the four horses on the Basilica di San Marco, brought by Doge Enrico Dandolo in 1204, and installed around 1254.

UK, London: From Charing Cross Rd, looking southeast to the northwest part of the front part of the English Anglican church St Martin in the Fields (1724, at the northeast corner of Trafalgar Square in the City of Westminster, spire height 59 m, 12 bells, tenor bell weight 1,486 kg, excavations under found a grave from about 410 AD (Roman era), in 1222 there was a church here, in 1542 Henry VIII rebuilt the church, in 1606 James I enlarged the church). It is famous for its regular lunchtime and evening concerts; Academy of St Martin-in-the-Fields performs here, and many other ensembles.

- Education Department

- Over 2 billions of children in the world are getting a solid peace-oriented education, to give a solid peace-oriented foundation for a good, free, peaceful and prosperous life.
- Education is very important – teachers work with parents and grandparents, to educate the children to leave healthy in a sustainable peace, liberty and prosperity.
- Discipline is strict, and those who do not behave properly, get medical assistance.
- The world has 4 school levels (SLs) of education:
SL1 – Kindergarten – 2 years: age 5 and 6
SL2 – Primary School – 4 years: age 7, 8, 9 and 10
SL3 – Secondary School – 3 years: age 11, 12 and 13
SL4 – High School or Vocational School – 4 years: age 14, 15, 16 and 17
- A World Library includes the Library of Congress and all the other great libraries – they remained where they are now, but are digitally interconnected, and accessible from any place in the world.
- adult education: technical, career
- training for employment
- management training
- post high school education
- peace education
- world constitution education

MARS: Without good education, there is no good future.

MOON: Exactly so – for this much attention is given to good education.
Aristotle: The roots of education are bitter, but the fruit is sweet.
Plato: Ignorance, the root and stem of all evil.

UK, London: The program by Candlelight at St Martin in the Fields, on Friday, October 14, 2016 at 7:30 PM, with Antonio Vivaldi (1678 in Venice-1741 (age 63) Vienna), Johann Sebastian Bach (1685-1750 (age 65)), Francesco Geminiani (1687-1762 (age 75)), George Frideric Handel (1685 Germany-1759 (age 74) London UK, buried at Westminster Abbey), Wolfgang Amadeus Mozart (1756 Salzburg, Austria-1791 (age 35) Vienna, Austria), Johann Pachelbel (1653 Nuremberg, Germany-1706 (age 53) Nuremberg, Germany).

- Science & Technology Department.

It helps in the areas of:
- mathematics
- statistics
- science
- technology
- Algorithmic Governance is an essential tool for a better and impartial governing of the world, used by the Advisers elected by people. Mathematicians from all countries work to improve the Algorithmic Governance, to better serve the people.
- cyberspace complete security was achieved long ago, and is strictly maintained
- information systems
- computer services
- Internet, AI
- scientific cooperation
- economic development at the world level
- infrastructure improvement and maintenance at the world level
- innovation and improvements in all areas, at the world level
- transportation at the world level
- safety
- security
- aviation
- highway
- cars
- railroads without noise
- maritime administration
- logistics
- strategic planning at the world level
- public works
- fleet maintenance
- standards: weights, measures, etc.
- research at the world level
- risk analysis
- laboratories

- engineering
- communications at the world level
- telecommunications
- networks
- peaceful nuclear energy use at the world level
- safety
- waste
- electrical power
- oceanic analysis at the world level
- atmospheric analysis at the global level
- meteorological service and prognosis at the global level
- world resources analysis
- sustainable use of world resources
- geographical and geological activity
- product safety at the global level
- hazardous material and chemical safety
- government broadcasting (radio, tv, Internet, newspaper, etc.) including news, scientific and technical information
- private broadcasting continues, but the world government must be able to directly inform the people, without intermediaries
- space exploration and expansion at the world level, including terraforming – very important for the future
- patent and trademark
- intellectual rights
- all government work, which can be done by private companies, is contracted with the best and reasonably priced private companies. At the same time, the government has competitive services for people – from plumbing and electrical help, to mortgage and buying or selling a house.

MARS: This new department has a large number of heavy tasks.

MOON: As expected, science and technology are becoming more important every day:
Pasteur: Science knows no country, because knowledge belongs to humanity, and is the torch which illuminates the world.
Galileo Galilei: In questions of science, the authority of a thousand is not worth the humble reasoning of a single individual.

UK, Oxford: From Broad St, looking south to the northeastern side of the Sheldonian Theatre (1669, classical concerts, lectures, ceremonies, capacity 1,000, by Gilbert Sheldon (1598-1677, Archbishop of Canterbury), 3 busts (center line), Bodleian Library (1602, left back, main research library of the University of Oxford, over 12 M items).

3543 Dialog 6. Moon terraforming

MARS: Well, what about you, Moon, and terraforming?

MOON: There is good work on terraforming me. As you know, there are many difficulties – for example the temperature fluctuations on my surface are ranging from minus 248 (everything is completely frozen) to plus 123 degrees Celsius (the water is boiling, etc.).
Now let's return for a moment on Earth.

The Advisers are elected every 20 months for one term only. If an Adviser X was elected for a term T1, then the next term T2 had another Advisor Y. For the next term T3, X can be elected again, but the next term T4 had a new Adviser, and so on. All levels of Advisers and other managers (minimum age 25 years, maximum 60) can be elected, not consecutively, at most 4 times (maximum 80 months = 6 years and 8 months).

All the employees in Government respect Seneca's aphorism: "To govern is to serve, not to rule", and Hippocrates' aphorism "Make a habit of two things: to help; or at least to do no harm."

MARS: Excellent implemented ideas.

MOON: Indeed:

Advisers have exceptional results obtained from their work, and based on these results, plus modesty, moderation, good character, friendliness, sharp mind, wisdom, good morals, and intense desire to help people, they are elected, without any campaigning, publicity, fundraising, donations, debates, propaganda, political parties, advertising, or similar activities.

UK, London: A beautiful Lloyds Bank (founded in Birmingham in 1765 by Taylors and Sampson Lloyd (1699-1779, iron manufacturer and banker)) building in London (founded by the Romans, who named it Londinium (a settlement established on the current site of the City of London around AD 43 (by Claudius (10 BC – 54 AD)). Its bridge over the River Thames turned the city into a road nexus and major port, serving as a major commercial center in Roman Britain until its abandonment around 420). London's ancient core, the City of London, largely retains its 2.9 km^2 medieval boundaries.

There is intense use of advanced digital technology, which opens up entirely new opportunities for developing direct elections, and public control of the institutions, improving the transparency of the election procedure, and taking into account the interests and opinions of each voter (over the age of 21, who are not in a special medical institution for bad behavior (which is a form of aggressive mental health) or for non-aggressive mental health).

MARS: The things are getting better.

MOON: Sure:
Thucydides: Ignorance is bold and knowledge reserved.

An Election Commission of 110 representatives from the 10 regions and from the 100 sub-regions, elected separately for 5 years, examine the qualifications of all the candidates for Advisers, and for other senior management positions. Unqualified candidates are asked to improve their qualifications, and then to try again later.

MARS: This is really necessary.

MOON: Right, we remember:
Otto von Bismarck: People never lie so much as before an election, during a war, or after a hunt.

UK, London: From the northeast corner of Trafalgar Square, south of the National Gallery, looking southwest to Vice Admiral Horatio Nelson's (1758-1805 (aged 47), buried at St Paul's Cathedral) Column, and the equestrian statue of King George IV (1762-1830 (aged 68), King 1820-1830, patron of architecture, the eldest son of King George III (1738-1820 (aged 81), Reign 1760-1820 (59 years), during his reign, the American colonies created the U. S. A.)).

It is important to refresh the management, and to bring new people to help the big family on the Planet. The older generations, who performed well, are retained in important roles, because experience and maturity count very much. At least two months before retirement, they are kindly asked to transfer their expertise to the younger generation. Even after retirement, they occasionally are invited to share their expertise.

In every election, with every winner, there are other two for number 2 and number 3. The number 2 and number 3 for each management position are used when number 1 is not available (vacation, sick, etc.). They constantly work for number 1, helping to solve urgent problems for the people.

Good elections are essential for the future.
Now on Earth there are clean and friendly elections, in which people choose between leaders with outstanding results, plus talent to lead people to peace and freedom, modesty, moderation, good character, friendliness, sharp mind, wisdom, good morals, and intense desire to help people – no campaigning, no publicity, no fundraising, no donations, no debates, no propaganda, no political parties, no advertising, or similar activities.

All Advisors are also local Administrators – they must show that they are good managers, and produce practical results for all people.

MARS: Yes, I notice that they are moving to better elections.

MOON: Indeed, they are refreshing the management, no propaganda, and they produce practical results for all people.
As Hoover said: Freedom is the open window through which pours the sunlight of the human spirit and human dignity.

Finland, Helsinki: the Railway Square, east of the railway station, with the Finnish National Theatre (1872 - 1902).

Japan, Kawaguchi, 17 km north-est of Mount Fuji (3776 m); a big Bonsai tree on the right and three smaller ones on the left.

3543 Dialog 7. Bio-robots for terraforming

MARS: I noticed a heavy use of bio-robots for terraforming.

MOON: Indeed, these multicellular robots are created by using biology and mechanical systems, and they are used in medical activities (drug delivery, healing, cancer recognition), for cleaning difficult areas, and, of course, for terraforming everything.

Now a short return on Earth.

An electronic world referendum is organized every three months. The main questions be:

1. Are you satisfied with the Government?
2. What Government work is good?
3. What Government work is not good?
4: Suggestions for improvement:

Within two months after each referendum, the Government responds to the people. Based on the suggestions received, new pro-people rules are replacing some old rules.

MARS: An electronic world referendum is refreshing.

MOON: Certainly, the people have the opportunity to express their opinions and suggestions.

MARS: When too many incorrect events take place, then what?

MOON: A change for the better becomes inevitable.

London, from the Shard (2012, 309 m, observatory at 244 m), looking east to the Tower Bridge (1886-1894, combined bascule and suspension turreted bridge over River Thames (flowing from west (left) to east (right)), between London boroughs Tower Hamlets (north – left up) and Southward (south – right), length 244 m, height 65 m, longest span 82 m, clearance 8 m (closed), 42 m (open)), City Hall (2002, height 45 m, center right round, for the Greater London Authority: Mayor of London and the London Assembly).

3543 Dialog 8. Deep space exploration finds many places ready for terraforming

MARS: I noticed many deep space explorations recently.

MOON: Yes, there is interest in exploring planets, moons, asteroids, comets, and other small bodies in our solar system, as well as beyond our solar system – there are fly-by missions, orbiters, and landings.
Now back to Earth for a moment.

Arms do not exist anymore, and only the police have some small arms. Those who want arms for hunting or sport, borrow them from police stations, with proper documents, rules and payments.

All military units have become strong civilian organizations, working to improve the quality of life for everybody.

MARS: An old dream has finally become realty.

MOON: Exactly so - we remember Edison: I am proud of the fact that I never invented weapons to kill.

UK, London: From a bus on Great George St, looking south to the north side of St Margaret's Church (1523), and in the back the eastern part of the north facade and entrance of Westminster Abbey (960, 1517, nave width 26 m, floor area 3,000 m^2, 2 towers, tower height 69 m, 10 bells, church of England, daily services, hosting all English and British coronations since 1066, and 16 royal weddings (from 1100 to 2011); it has the tomb of King Henry III of England (1207-1272).

3543 Dialog 9. Multi-loop coupling mechanisms for terraforming

MARS: There are many types of tools used for terraforming.

MOON: Indeed, for example multi-loop coupling mechanisms (MCMs), which are complex spatial mechanisms with high rigidity, large working space, and at least one coupled constraint limb in the chains that connect the base link and the output link.
 Let's go to Earth for a second or two.

A census takes place every 5 years, and all people have received a special credit card (SCC), with their photo and other personal data. The delimitations between regions, and between sub-regions, are adjusted by the census.

MARS: Simple and elegant.

MOON: Every 5 years we'll know better the world population.

Italy, Venezia - The south end of La Piazzetta, the south part of Piazza San Marco, with gondole, and a wedding picture of a Japanese couple.

UK, London: From the northeast corner of Trafalgar Square looking west to the southeast façade of The National Gallery (1824, 2,300 paintings).

Finland, Helsinki: a tall ship in the tourist harbor, in the south-east part of the city.

3543 Dialog 10. Meteors and terraforming

MARS: Meteors are a big challenge for terraforming.

MOON: Yes, there are over 25 millions of meteors of different dimensions which enter just Earth's atmosphere each day – they add about 15,000 tons of material on Earth each year. Where there is no atmosphere, protection against meteors is a high priority, because there are over 4.5 billions of meteors in the solar system.
Now back to Earth for a moment.

The special credit card (SCC) are used to buy everything, to identify for voting, for census, for travel, for medical assistance, etc.
There are also private credit cards, which are very competitive.
The changes of the delimitations between regions, and also sub-regions, is easily inputted on these cards, and no other work is needed.

MARS: Important idea.

MOON: The special credit card (SCC) are much more than a credit card, and simplify life for everybody.

MARS: Tempora mutantur, et nos mutamur in illis.

MOON: The times are changing, and we are changing with them.

Rome, Vatican, Piazza San Pietro (1667, by Gian Lorenzo Bernini): Basilica di San Pietro (1506, center back), granite fountain by Carlo Maderno (1614, center, north side of piazza).

Finland, Helsinki: the square near a harbor, with the Presidential Palace, the City Hall and other government buildings on the right

3543 Dialog 11. Places for terraforming

MARS: Any idea how many places are out there for terraforming?

MOON: Well, there are over 1,000,000 asteroids with a diameter of over 1 km (surface area over 3.14 km^2), and many of them are really good for terraforming.
   Let's see what is going on Earth.

It is well known for long time that the enemies of the people on Earth are not other people, but viruses, microbes, bad bacteria and hundreds of deadly illnesses – all people on Earth work together against these real enemies for all.

MARS: What a beautiful idea?

MOON: I love it – of course people are something sacred for people, and all people on Earth work together against viruses, microbes, bad bacteria and hundreds of deadly illnesses.

MARS: Tell me again what the people really want.

MOON: All people on the planet want peace, disarmament, freedom & order, good health, good education, good economy, and prosperity for all.

Japan, Mount Fuji (Fuji-san, 3776 m), seen from Fujiyoshida, circa 15 km north-est from Mount Fuji, 1000 m altitude.

Rome: residential apartments near Via Aurelia (constructed around 241 BC by C. Aurelius Cotta, who was a censor) and Piazza Irnerio, about 9 km east of Trajan's Column (113 AD).

3543 Dialog 12. Ceres for terraforming

MARS: Is Ceres interesting?

MOON: Certainly, Ceres is the largest asteroid, about 25% of my size, diameter of about 914 km, surface area about 2,620,000 km^2, and orbits the Sun between Mars and Jupiter, where there is the asteroid belt.
    Back to Earth for a moment.

    Non-violence is a strict requirement for all activities on Earth.
    The first rule for everybody on Earth comes from the Hippocratic Oath: Primum non nocere - first do not harm.
    Medical doctors and assistants make regular home visits to all people, to keep them healthy, and to prevent illnesses.

MARS: This is fundamental. Remember Edison?

MOON: Sure: "Non-violence leads to the highest ethics, which is the goal of all evolution."

UK, Cambridge: From Trinity Lane looking south to the west part of the northern façade and entrance of King's College Chapel (1446, center back, the College was founded in 1441, and the Old Schools was part of King's College), the east gate of Clare College (1326, as University Hall, making it the second-oldest college of the University, after Peterhouse (1284)) and its Chapel (1763, center right), and the Old Schools (1441, University Offices, left).

3543 Dialog 13. Vesta for terraforming

MARS: What about Vesta?

MOON: I am glad that you asked – Vesta is ready for terraforming, has a dimeter of about 525 km, surface area 866,000 km^2, 2.3 times farther for the Sun than the Earth, and looks attractive.
    Earth is waiting:

    People need only truth in order to create a long term peaceful and harmonious society.
    If someone lies – medical treatment follows.

MARS: Hard work was needed to have again truth.

MOON: True, but, step by step, with calm and patience, the truth is back, simply because it is much better than any lie.
Giordano Bruno: Truth does not change because it is, or is not, believed by a majority of the people.
Cicero: Nature has planted in our minds an insatiable longing to see the truth.
Cicero: Faithfulness and truth are the most sacred excellences and endowments of the human mind.
Goethe: Wisdom is found only in truth.
Rousseau: Falsehood has an infinity of combinations, but truth has only one mode of being.
Aeschylus: In war, truth is the first casualty.
Churchill: A lie gets halfway around the world before the truth has a chance to get its pants on.

Italy, Venezia - Piazza San Marco with Palazzo Ducale (right), Libreria Sansoviniana (next to Palazzo Ducale), Basilica di San Marco (back), Giardini Reali and Il Campanile (center-right), Procuratie Nuove (center to left), Capitano di Porto (left).

3543 Dialog 14. Pallas ready for terraforming

MARS: Indeed, this B-type asteroid is spectrally blue.

MOON: Yes, and Palla has a diameter of about 511 km, surface area about 820,000 km^2, and is about 414.8 M km from the Sun.
Back to Earth for a second.

Freedom is a fundamental requirement on Earth.
It is well understood that this freedom refers to doing good things in a civilized manner, not for war, violence or similar bad things, which are against the wellbeing of the people.
Freedom goes hand in hand with responsibility.
People can assemble peacefully only.
For economy it is clear that the free market economy, while not perfect, gives the best results, but all people have the option to choose between friendly private services, and friendly government services. Independent assistants and monitors every day make sure that there are no abuses. Sine qua non requirements for happiness are morality and free market.
Religion is free, and helps people.
People of course can petition the small Word Government, and can change it anytime, if it does not perform as expected.

MARS: Not so easy job…

MOON: Sure, they need calm, patience, constant discussions, and wise leaders, with the courage to change some old habits, and move happily to freedom, free markets, and people-oriented decisions.
Thucydides: The secret to happiness is freedom… And the secret to freedom is courage.
Epictetus: Freedom is not procured by a full enjoyment of what is desired, but by controlling the desire.

Finland, Helsinki: the Railway Square, east of the railway station, the bus station (left) and the Finnish National Theatre (center-left).

3543 Dialog 15. Hygiea can't wait terraforming

MARS: This C-type asteroid, which is the most common, is attractive indeed.

MOON: Yes, Hygiea has a diameter of about 433 km, surface area of about 589,000 km^2, and a distance from Sun of about 469.588 M km.

Let's relax on Earth.

All budgets have a surplus of 2% - there is a strict application of the Latin aphorism: "Sumptus censum ne superset" (Let not your spending exceed your income).

MARS: That's nice.

MOON: Yes, not having any war-related expenses, it is much easier to balance the world budgets, and even have a surplus of 2%.

3543 Dialog 16. Interamnia is ready for terraforming

MARS: This large F-type asteroid was discovered by Vincenzo Cerulli on 2 October 1910, in Teramo, Italy, and he named it after the Latin name for Teramo.

MOON: Interesting, Interamnia has a diameter of about 332 km, a surface area of about 346,000 km^2, and is about 458 M km from the Sun.
   The Earth has news:

   Correcting errors is a permanent duty for everybody - Darwin said "To kill an error is as good a service as, and sometimes even better than, the establishing of a new truth or fact."

MARS: As you know, the world was full of errors.

MOON: Yes, they accumulated over many years, and a good effort was necessary to correct them all, but it was done relatively fast.

Japan, Nagoya: a tall building seen from Shinkansen (the bullet train, 320 km/h, started in 1964),

3543 Dialog 17. Europa wants fast terraforming

MARS: This sixth largest asteroid was discovered on 4 February 1858 by Herman Goldschmidt from his balcony in Paris, and named after one of Zeus's conquests in Greek mythology – Jupiter's smallest of the four Galilean moons is also named Europa.

MOON: Fascinating – Europa has a diameter of about 319 km, surface area is about 320,000 km^2, and is 463 M km from the Sun.
    Let's go to Europa on Earth.

    Kindness is a requirement for everybody.
    Seneca said "Wherever there is a human being, there is an opportunity for a kindness."
    This is a fundamental idea which is constantly applied in Europa and everywhere.

MARS: Seneca said it right.

MOON: And they all follow his good advice, because, clearly, it is for the benefit of all of them.

Finland, Helsinki: in the south of the Railway Square is the Ateneum (1887, a major museum of classical art).

3543 Dialog 18. Sylvia graciously asks for terraforming

MARS: Sure, Sylvia is a X-group asteroid, discovered on 16 May 1866.

MOON: Yes, Sylvia has a diameter of about 271 km, surface area of about 231,000 km^2, and distance to the Sun of around 521.3 M km.
Now Earth needs some attention:

All levels of government are highly mobile – they change often capitals for the 10 regions, and for the 100 sub-regions, etc.
It is necessary to move the government close to the people, to be able to quickly solve the local problems.
Locally the people decide how to better organize themselves, to be more efficient and harmonious, with the help of the world government when necessary. Like in any big family, there are differences in organization and management, based on their abilities and objectives, but all must be peaceful and harmonious. Conflicts are promptly resolved by the medical personnel, police, and other assistants.

MARS: Mobility is actually a pleasant requirement.

MOON: Certainly, changing capitals, meeting new people, are nice activities, and also help improve the quality of life for all.

MARS: Remember Confucius?

MOON: The more man meditates upon good thoughts, the better will be his world, and the world at large.

UK, Cambridge: From Trinity Lane looking southeast to the west façade with the entrance to the Old Schools (1441, University Offices, the administrative center of the university, surrounded to the north by Gonville and Caius College (1348), to the east by the University of Cambridge Senate House (1722, where degree ceremonies are held, on King's Parade), to the south by the King's College Chapel (1446), and to the west by Trinity Hall (1350) and Clare College (1326)).

3543 Dialog 19. Ceramics for terraforming

MARS: Ceramics are very useful in the difficult environment of space.

MOON: Indeed, they are commonly used for terraforming, and heat shields, to protect the spacecraft and its passengers. Alumina, one of many types of ceramics, is used in bearings for its wear resistance and electrical insulation purposes in spacecrafts. Since weight is very important in the space, ceramics can be lighter while maintaining strength compared to metals.
Now back to Earth.

The United Nations was very useful, and was changed into World Police and Assistance Organization (WPAO), to help local police in case of big natural disasters or big accidents, and report to the top 10 Advisers. They are located in all capitals, and help the locals. When an emergency appears, they quickly move to solve the emergency.
The police powers are limited, and they know and are friendly with all the people in their jurisdiction – this is the key element of a civilized and peaceful Earth.
Police are people's friends everywhere, and they always help people.
Prevention of bad events is the main objective of everybody. If a bad event occurs, the police and their assistants eliminate the consequences, reestablish the normal situation, and determine why the bad event occurred, in order to improve their activity, and prevent such bad events in the future.
Private property cannot be taken for public use, without just compensation, decided by at least 5 assistants.
A person cannot be deprived by government of life, liberty, or property, without having several doctors and other assistants agree: for life – at least 12; for liberty – at least 6; for property – at least 3.
A person cannot deprive another person of life, liberty, or property, which, unfortunately, was very frequently in the world,

and very much effort and energy was allocated to prevent such bad events.

In order to prevent bad things, the police, doctors and their assistants are in permanent contact with all the people, by visiting them, phone calls, e-mails, tele-videos, and mail, to keep everybody calm and happy.

MARS: These are serious and demanding issues.

MOON: Yes.
Solon: In giving advice seek to help, not to please, your friend.

Italy, Venezia - The south façade of Basilica Cattedrale Patriarcale di San Marco, with three of the five domes visible up right.

Finland, Helsinki: on the west side of the Railway Square there is this beautiful building called Fennia, built in 1899, which was first a hotel, then a restaurant, now is the Casino Grand Helsinki.

3543 Dialog 20. Eunomia wants terraforming now

MARS: This asteroid is the largest of the S-type (stony), and may contain about 1% of the mass of the asteroid belt.

MOON: Impressive, Eunomia has a diameter of about 270 km, a surface area of about 229,000 km^2, and the distance to the Sun is about 400 M km.
 Earth is online:

 About 66% of the people of the world are working at any moment. Therefore, non-stop working of all world government departments – especially medical, police, emergency, volunteers – are carefully organized.

MARS: No question that essential government departments are non-stop working.

MOON: Sure, in fact all the departments have employees working anytime somewhere in the world, but the essential government departments are in all places non-stop working.

MARS: If people have questions, what do they do?

MOON: First ask around, maybe somebody can respond. If not, the government has several phones and e-mails for questions, and in 3 days they send a response. This is fundamental – the government always is ready to respond and help people.
All the public work is Not for me, not for you, but for us – from Latin: Non mihi, non tibi, sed nobis.

3543 Dialog 21. Euphrosyne likes terraforming

MARS: This is a young asteroid, named after one of the Charities in Greek mythology.

MOON: I see, Euphrosyne has a diameter of about 268 km, surface area of about 226,000 km ^{2}, and distance to the Sun is about 471 M km.
    Back to Earth:
    In order to have serious and constructive discussions and negotiations, they must be private.
    Privacy and discipline are necessary for good government work.
    The results are public and preserved, but not the private discussions.

MARS: Without privacy, not much can be accomplished.

MOON: Indeed, private serious discussions, with fast and good results, are the norm.
Vegetius: Few men are born brave. Many become so through training and force of discipline.

3543 Dialog 22. Cybele is on the list for terraforming

MARS: This asteroid appears to be a remnant primordial body, and was named after the Earth goddess.

MOON: Yes, Cybele has a diameter of about 263 km, a surface area of about 217,000 km^2, and the distance to the Sun about 514.4 M km.
   Listen to the Earth now:
   It is a strict requirement for the top management, and for all others, to be highly civilized, polite, courteous, harmonious and efficient.
   Whoever wants to work for the world government must have good manners.
   Harmony in the world starts from the harmony and good manners of the people in the world government.

MARS: Beautiful requirement…

MOON: The government must be an example in good manners: highly civilized, polite, courteous, harmonious and efficient.
Churchill: It`s not enough that we do our best; sometimes we have to do what`s required.

UK, Oxford: On St Aldate's St, 140 m north of Tom Tower, the south side and entrance (right) of the Museum of Oxford (1975, in the former premises of the Oxford Public Library), a history museum of the City and University of Oxford, from prehistoric times onwards, with original artifacts, Roman pottery, details about King Charles I of England (1600-1649, king 1625-1649, who had Oxford as his stronghold), Oliver Cromwell (1599-1658), etc.

3543 Dialog 23. Juno is longing for terraforming

MARS: This asteroid was the third to be discovered on 1 September 1804, by Karl Harding, and is S-type (stony).

MOON: Nice, Juno has a diameter of about 254 km, a surface area of about 200,000 km^2, and the distance to the Sun is about 400 M km.
   Earth, please:
   All conflicts are not only quickly resolved, but they are transformed in friendships. This is very important for long term stability.
   The medical personnel and others work diligently to make sure that disputes are resolved, and then a friendship is developed. Only in this way the situation becomes stable.
   People want peace, freedom, health, friendship and prosperity, therefore conflicts are quickly resolved, and then the corrective medical treatment includes the transformation of hostility and aggressiveness into harmony and friendship.

MARS: This is a demanding task.

MOON: Right, but it is done by talented doctors, who work with patients until the results are satisfactory.
   Newton said: Men build too many walls and not enough bridges.

Finland, Helsinki: the square near the passenger harbor (left), with the Presidential Palace, the City Hall and other government buildings on the right.

Japan, Mount Fuji (3776 m) seen from 12 km north-east, on Higashi-Fuji-Goko Road, in Fujiyoshida; the road 701 to Fuji

3543 Dialog 24. Patentia patiently waits for terraforming

MARS: This asteroid was discovered on 4 December 1899 by Auguste Charlois, and has an orbital period of 5.36 years.

MOON: Great, Patentia has a diameter like Juno, of about 254 km, a surface area of about 200,000 km^2, and the distance to the Sun is about 457.6 M km.
   Back to Earth.
   As a single big, over 8 B, family on Earth, all people must be able to communicate easily with each other.
   For this reason, a common language and alphabet on Earth are needed. Because English was a de facto common language long ago, it was taken as the basis of the world language, called Mundo, which is taught in all schools, and used in the world government. All the other languages continue as secondary languages.
   The same is true for the Latin alphabet, which is used everywhere, with other alphabets as secondary.
   The teachers had a very significant role in implementing this task.

MARS: Easy communication requires very hard work.

MOON: Absolutely, but it is very important. You see, having hundreds of languages and dialects, and several alphabets, does not help to come together as a big family.

UK, Oxford: On Merton Street an entrance to Corpus Christy College (1517, founder Richard Foxe, the Bishop of Winchester, 12th oldest college in Oxford (1st University College (1249, 2nd Balliol College (1263), 3rd Merton College (1264)), 249 undergraduates, 94 postgraduates), situated between Merton College (1264, founded by Walter de Merton (1205-1277), Lord Chancellor to Henry III (1207-1272) and later to Edward I (1239-1307), and Catholic Bishop of Rochester (1274-1277); Merton College Library (1373) is the oldest functioning library in the world), and Oriel College (1326).

3543 Dialog 25. Bamberga loves terraforming

MARS: This asteroid was discovered on 25 February 1892 by Johann Palisa, in Vienna, and is a C-type.

MOON: I see, Bamberga has a diameter of about 227 km, a surface area of about 162,000 km^2, and the distance to the Sun is about 400 M km.
    Now this is Earth.

    Like in any big family, there are differences, because some work more, some spend less, some move faster, and, especially, some are sick – this is the main reason for differences: not all people can be equally sick, some people are sicker than others. However, all the people and the government work to help each other.
    It is a major responsibility of the Government to increase global wealth, and to train those in need to have better working abilities and opportunities.

MARS: Really good points we have here.

MOON: Yes, wealth is important, all the people and the government work to help each other.

Italy, Venezia - Palazzo Dandolo on Riva degli Schiavoni, 150 m east of Piazza San Marco.

3543 Dialog 26. Psyche is looking for terraforming

MARS: This asteroid was discovered on 17 March 1852 by Annibale de Gasparis, and named after the Greek goddess Psyche.

MOON: Good, Psyche has a diameter of about 223 km, a surface area of about 156,000 km^2, and the distance to the Sun is about 463.8 M km.
    Now our beloved Earth:

    No bureaucracy – this is required by all people, and every day attention is given for improvements in this direction.
    In a well-organized country, with all people working together in harmony, this can be easily accomplished.

MARS: No bureaucracy looks like a nice dream.

MOON: They want this dream to become reality – not over night, but working every day, for many years, with patience and perseverance, and it was accomplished.
Reagan: No government ever voluntarily reduces itself in size. Government programs, once launched, never disappear. Actually, a government bureau is the nearest thing to eternal life we'll ever see on this Earth!

Japan, Lake Kawaguchi (17 km north-est of Mount Fuji), 833 m altitude, 15 m depth, 19 km shore length, 6.13 km^2, in the morning

3543 Dialog 27. Thisbe – to be or not to be terraformed

MARS: This asteroid was discovered on 15 June 1866 by C. H. F. Peters, was named after a heroin of a Roman fable, and has a period of 4.6 years.

MOON: Beautiful, Thisbe has a diameter of about 218 km, a surface area of about 150,000 km^2, and the distance to the Sun is about 781 M km.
Earth, what's up?

Well, everybody works really hard to completely eliminate corruption, organized crime and drug trafficking.
Constant attention is focused on avoiding duplication at all levels of the world government – there must be continuous collaboration between all levels, to prevent duplication, and to eliminate it, if it was found.

MARS: All these are big tasks.

MOON: Certainly, but calm, friendly and well-organized effort, involving also medical personnel and police, finally eliminated these problems. The idea is that nothing is impossible, if the people really want to do it.

Washington, D.C. (1790): George Washington (1732-1799, first President 1789-1797) Monument (1848-1885, 169 m, 43 ha), on the National Mall, 700 m south of the White House, seen from the Constitution Avenue NW.

Italy, Rome. Isola Tiberina in the middle of Tiber river, which flows to the back, viewed from Ponte Garibaldi (1888), with Pons Cestius (27 BC, the first stone bridge, right, connecting Isola Tiberina with Trastevere).

3543 Dialog 28. Silicones for terraforming

MARS: It is well known that outer space missions and exploration for terraforming present extraordinary challenges in space craft design, which include broad temperature cycling, outgassing, atomic oxygen resistance, UV radiation, weight restrictions, and vibration

MOON: Indeed, and silicones have been a key material choice for aerospace and terraforming.
 Back on Terra for a moment.

 Each government department have some reserves for special situations (natural disasters, big accidents), and the banks also have good financial reserves.
 All people are encouraged to save some money in banks with 5% interest.

MARS: Reserves and savings are obviously important.

MOON: Yes, and here there are some changes – you see, everybody needs to have some government guaranteed savings in banks, with 5% interest.

MARS: Vitam mutaveris in meliores actus.

MOON: Change your life for the better.

Italy, Venezia - In the middle of the west façade of the Basilica di San Marco, we see the central bronze-fashioned door, in a round-arched portal, encircled by polychrome marble columns. Above this door there are three round bas-relief cycles of Romanesque art. A Japanese couple, with their Japanese photographer, make their wedding photographs in this most beautiful place.

3543 Dialog 29. Doris is thinking terraforming

MARS: This asteroid was discovered on 19 September 1857 by Hermann Goldschmidt from his balcony in Paris, was named after a Oceanid in Greek mythology, and passed within 2.8 M km of Pallas in June 2132.

MOON: That's really close - Doris has a diameter of about 215 km, a surface area of about 145,000 km^2, and the distance to the Sun is about 465 M km.
    Earth, your turn:

    Thank you - inspectors help the Government with integrity and efficiency issues – always there are ways to improve the work.
    Inspectors give advice regarding integrity and efficiency, and take corrective actions when necessary.

MARS: Always there is room for improvement in these areas.

MOON: Sure, for this the friendly inspectors help with advices and big smiles.

Finland, Helsinki: The Three smiths statue (by Felix Nylund, 1932), with the Old Student House (1870, left) and Tallberg's house (right). On the base: MONUMENTUM – CURAVIT – LEGATUM – J. TALLBERGIANUM – PRO HELSINGFORS A.D. MCMXXXII ("The statue was erected with the help of a donation from J. Tallberg by Pro Helsingfors in the year 1932").

UK, Oxford: From Broad St, looking southeast to the north façade of Clarendon Building, the registration ceremony at the University of Oxford.

3543 Dialog 30. Fortuna needs assistance for terraforming

MARS: This asteroid was discovered on 22 August 1852 by J. R. Hind, and has a composition similar to Ceres.

MOON: I see - Fortuna has a diameter of about 211 km, a surface area of about 140,000 km^2, and the distance to the Sun is about 365 M km.
  Back to Earth for a second:

  Because all families need assistance from time to time, and the big family on Earth contains billions of small families, all of them now have the assistance they need – this be the result of one country well organized and managed.

MARS: Remember Homer?

MOON: Certainly: There is nothing nobler or more admirable than when a man and a woman, who see eye to eye, keep house as man and wife, confounding their enemies, and delighting their friends.

Japan, Mount Fuji (Fuji-san, 3776 m), seen in the morning from the window of a hotel in Kawaguchi, circa 17 km north-est.

Finland, Helsinki: the north-east side of the Railway Square, with the Radisson Blu Plaza Hotel Helsinki (center-left) and the Casino Helsinki (right). In winter time, the Railway Square hosts an ice-skating rink and the people from Casino and from hotel enjoy skating just near the door. It seems that the earliest ice skating happened in southern Finland, maybe around this place, more than 3000 years ago.

3543 Dialog 31. Themis is optimistic about terraforming

MARS: This asteroid was discovered on 5 April 1853 by Annibale di Gasparis, and was named after Themis – the personification of natural law and divine order in Greek mythology.

MOON: Nice indeed - Themis has a diameter of about 208 km, a surface area of about 136,000 km^2, and the distance to the Sun is about 470 M km.
    Look at Earth for a moment:

    Because all people on Terra wanted to live in harmony for millennia, it was relatively easy to implement this harmony in one good and civilized country.

MARS: Harmony, like in music, is so important…

MOON: And they finally have it.
    The first rule for everybody on Earth comes from the Hippocratic Oath: Primum non nocere - first do not harm.

3543 Dialog 32. Aurora graciously waits for terraforming

MARS: This asteroid was discovered on 6 September 1867 by J. C. Watson, and was named after the Roman goddess of the down (which starts when the center of the Sun is 18^0 below the observer's horizon, and ends at sunrise).

MOON: Fascinating - Aurora has a diameter of about 205 km, a surface area of about 132,000 km^2, and the distance to the Sun is about 473 M km.
  Back to Earth – is waiting….:

  Dispute resolution is not only Government's obligation, but it is everybody's duty.
  There is professional assistance from medical personnel, police, people assistance specialists, volunteers, religious organizations, and many others, but the bottom line is that everybody must avoid disputes.

MARS: Avoiding disputes is a good objective.

MOON: Sure, nobody is right all the time, therefore everybody accepts new and constructive ideas, which have practical benefits for all.
To err is human, to persevere is of the devil – from Latin:
Errare humanum est, perseverare diabolicum.

Finland, Helsinki: trams (first appeared in 1807 in UK) on Mannerheimintie.

Rome. Down: a part of Forum Augustum (2 BC). Back: a part of Forum Traiani (113 AD), including the Columna Traiani (113, center back).

3543 Dialog 33. Amphitrite smiles to terraforming

MARS: This asteroid was discovered on 1 March 1854 by Albert Marth, and was named after a sea goddess in Greek mythology.

MOON: Interesting - Amphitrite has a diameter of about 204 km, a surface area of about 131,000 km^2, and the distance to the Sun is about 382 M km.
Let's see the Earth now:

Special attention is given by Advisors to avoid abuses and wrong interpretations of the rules. All assistants (doctors, mathematicians, CEOs, engineers and teachers) closely monitor all activities, to avoid abuses and wrong interpretations of the rules.

This requirement of not having abuses is demanding – but this is a general job, not only for Government, but for everybody, as part of the big family, they just don't need abuses.

The abuse, in some places, of confiscating the land by some government bureaucrats was eliminated – the land belongs to the people, not the government.

The abuse, in some places, of having trains, airplanes, and others making unhealthy noises, with the government support, was eliminated – peoples' health has always priority.

The abuse, in some places, of having to change the clocks twice a year was eliminated – only the normal local time zones are used.

If abuses are observed, they are immediately reported to the Government, and corrected, in general, by the People Assistance Department, which has personnel, including medical assistants, to analyze and promptly solve the abuses.

MARS: Abuses were frequent…

MOON: Yes, unfortunately, and abuses require some extra attention from the government, and from all people. The abuse of naming hurricanes with people's names was eliminated, wasting people's money on unnecessary projects was eliminated, and many others. Step by step, it was done.

USA, New York (1624): on 42nd street, close to 8th Avenue, inside a tall building, three sculptures of people waiting at a door.

Japan, Kyoto, Kyoto Century Hotel, 10 km north-west of Byodo-in Temple (998), seen from Shinkansen (the bullet train, 320 km/h).

3543 Dialog 34. Egeria is eager for terraforming

MARS: This asteroid was discovered on 2 November 1850 by Annibale de Gasparis, and was named after the mythological nymph Egeria of Aricia, Italy, the wife of Numa Pompilius, second king of Rome.

MOON: Charming - Egeria has a diameter of about 202 km, a surface area of about 128,000 km^2, and the distance to the Sun is about 385 M km.

A short break for Earth:

In one country, with one market, the commerce between the people on Earth is free of taxes, tariffs, duties, etc. – plenty of opportunities for everybody.

The speech is free and responsible. It is expected not to call for war, violence, or similar destructive activities. People want peace, freedom, health, friendship and prosperity.

The press is free and responsible. It is expected not to call for war, violence, or similar destructive activities. People want peace, freedom, health, friendship and prosperity.

People can assemble peacefully only, with police for help. It is expected not to call for war, violence, or similar destructive activities. People want peace, freedom, health, friendship and prosperity.

MARS: Freedom, of course, goes hand in hand with responsibility.

MOON: Exactly, and the most important freedom is to quietly elect new world government leaders every 20 months, and to have referendums on their work every 3 months.

MARS: Cicero said it nicely….

MOON: Freedom is a possession of inestimable value.

Finland, Helsinki: commercial buildings south (left) and west (center) of Helsinki Central Railway Station (1907 – 1914).

UK, Oxford: On Merton St. at Magpie Lane (to right, to Old Bank Hotel), looking west to the south part of Oriel College (1326).

3543 Dialog 35. Elektra wants electricity with terraforming

MARS: This asteroid was discovered on 17 February 1873 by C.H. F. Peters, has 3 moons, and was named after an retaliator in Greek mythology.

MOON: 3 moons! - Electra has a diameter of about 199 km, a surface area of about 124,000 km^2, and the distance to the Sun is about 3.4 B km.
　　Look to Earth:

　　There are always plenty of jobs (assisting other people, for example), and the standard situation is this: more jobs than available people, so people choose the jobs they like the most.
　　No unemployment, no homelessness, no begging, no tipping – just all working harmoniously, having good houses, and helping each other.

MARS: This is definitely a big success.

MOON: Certainly, everybody has a free e-mail from the government, in addition to private e-mails, to be able to freely communicate, find jobs, etc., without any advertisements. Having a job which you like is really important.

UK, Cambridge: From Trinity Ln, looking west through the entrance of Trinity Hall, (1350, by Ian Baterman (c 1298-1355, Bishop of Norwich between 1344 and 1355), a constituent college (the 5th oldest) of the University of Cambridge), to the Front Court and the entrance to the west building of the Front Court. To the northeast of Trinity Hall there is the separate Trinity College (1546, founder Henry VIII (1491-1547, reign 1509-1547), motto: Virtus Vera Nobilitas).

3543 Dialog 36. Iris is longing for terraforming

MARS: This asteroid was 7th discovered on 13 August 1847 by J. R. Hind, is the fourth-brightest in the asteroid belt, is an S-type (stony) and was named after the rainbow goddess Iris in Greek mythology, who was a messenger to the gods, especially Hera – the Roman equivalent of Hera is Juno, which is the 3rd discovered asteroid.

MOON: Charming - Iris has a diameter of about 199 km, a surface area of about 124,000 km^2, and the distance to the Sun is about 825.78 M km.
    Let's stop to see Earth:

    All standards or rules proposed by Advisers must be approved by their 5 assistants (doctors, mathematicians, CEOs, engineers and teachers), and for any new rule over 2,000 basic rules (each rule on at most half a page, total 1,000 pages), at least on old rule must be eliminated.
    All the rules can be changed or eliminated when a majority of the people or their Advisors agree, but some fundamental peace and order rules remain.

MARS: Good and refreshing idea.

MOON: Actually, after a number of years, when all the people on the planet are highly civilized, friendly and respectful, no rules are needed, because everybody knows from schools and from experience, how to comport themselves, and how to work together for the benefit of all.

3543 Dialog 37. Hebe

MARS: This asteroid was the 6th discovered on 1 July 1847 by Karl Ludwig Hencke, contains about 0.5% of the mass of the belt, has high bulk density (greater than that of the Moon), solid body, is the fifth-brightest object in the asteroid belt after Vesta, Ceres, Iris, and Pallasand, and was named after the Greek goddess of youth, name proposed, at Hencke's request, by Carl Friedrich Gauss (30 April 1777 – 23 February 1855, aged 77.8 years, a very important mathematician, astronomer, geodesist, and physicist who significantly contributed to many fields in mathematics and science. He was director of the Göttingen Observatory and professor of astronomy for over 47 years, from 1807 (age 30) until his death in 1855 (age 77.8)).

MOON: Fascinating - Hebe has a diameter of about 195 km, a surface area of about 119,000 km^2, and the distance to the Sun is about 2.2 B km.
 Back to Gauss' Earth:

 The Constitution of the World can be improved when 66% of the voters agree.

MARS: The Constitution of the World is very good indeed, but always there is room for improvement.

MOON: After much more experience is accumulated as a single country on the planet, there are new ideas to be added, which make the life even better for all.

UK, Oxford: From the main entrance, looking east to the east side of the Front quad (the oldest collegiate quadrangle) of Merton College (1264)

3543 Dialog 38. Eugenia

MARS: This asteroid was the 45th discovered on 27 June 1857 by the amateur astronomer Hermann Goldschmidt, has a Moon, and was named after the wife of Napoleon III – the first asteroid not named from classical mythology..

MOON: I see - Eugenia has a diameter of about 188 km, a surface area of about 111,000 km^2, and the distance to the Sun is about 988.84 M km.
   Look at the Earth – is waiting….:

   The purpose for all people on Terra is to be healthy, to live in peace, freedom and harmony, to be prosperous, and to prepare to expand to the Moon, asteroids, Mars, and other places in the Universe, which can support life.

MARS: This purpose is excellent.

MOON: I can't wait to get there! Especially having people from Terra living on me, the Moon, on asteroids, on you, Mars, and other places in the Universe, which can support life, is absolutely adorable!

   Important immediate objectives for everybody on Terra are:
- Reserve time for happiness.
- Use robots and automated processes, work less, and spend more time with your family.
- The weekend is like a small vacation.
- Prevent burnout.
- Make civilized behavior and harmony everywhere an important issue.
- Eliminate stress.
- Help friends and colleagues.
- Keep everybody relaxed, calm, friendly, patient, and happy.

For better understanding and easier implementation, the following books, by Michael M. Dediu, are recommended:
- Our Future is Sustainable Peace and Prosperity – Moving from conflicts to harmony and peace
– Our Future Depends on Good World Educations – Moving from frail education to solid education.
– Friendly, Helpful & Smart World Management - Moving from bureaucracy to responsive world management
– If You Want Peace, Prepare for Peace! – Moving from preparation for war to preparation for peace
– World with One Country & its Ten Friendly Regions - Moving from 195 disagreeing countries, to 1 country with 10 collaborating regions
– After 10,000 Years of Conflicts, People want 10,000 Years of Harmony - Moving from continuous wars to stable peace
- World Constitution Implementation – Moving from violent changes, to smooth transition to the Constitution of the World
- It is getting truer and truer – we urgently need the World Constitution: Moving from anarchic changes, to balanced transition to the Constitution of the World

There are also over 90 World Monthly Reports, with very interesting and useful information.

Many more books are in the Bibliography.

UK, Oxford: From Merton St. looking south to the northern façade of the main entrance of Merton College (1264). Important personalities associated with Merton College are British chemist Frederick Soddy (1877-1956, Nobel Prize in Chemistry (1921)), poet T. S. Elliot (1888 in St Louis, U. S. – 1965 in London, England, Nobel Prize in Literature (1948)), British philosopher John R. Lucas (born 1932), British mathematician Sir Andrew Wiles (born 1953, proved Fermat's (1607-1665) Last Theorem (1637) proved after 358 years).

3543 Dialog 39. Metis is patiently waiting for terraforming

MARS: This asteroid was discovered on 25 April 1848 by Andrew Graham, passed about 5,000,000 km of Vesta on 19 August 2004, and was named after the Greek mythological Metis, a Titaness and Oceanid, daughter of Tethys and Oceanua.

MOON: Great - Metis has a diameter of about 173 km, a surface area of about 94,000 km^2, and the distance to the Sun is about 834.76 M km.
　　　Now let's see the Earth:

　　　This Constitution of the World is valid not only on Terra, but also on the space around Terra, on the Moon, Mars, asteroids and any other places were the very good people on Terra are moving, and will be moving in the future.

MARS: This is a big joy – peace everywhere!

MOON: And freedom, and prosperity – what do you want more?!

USA, Boston (founded in 1630): visiting tall ships from many countries, at the Boston Fish Pier (opened in 1915).

3543 Dialog 40. Eleonora is euphoric about terraforming

MARS: This asteroid was the 354th discovered on 17 January 1893 in Nice, by August Charlois, is an S-type (stony), and the spectrum of Eleonora shows the strong presence of the mineral olivine, a relative rarity in the asteroid belt.

MOON: Interesting - Eleonora has a diameter of about 165 km, a surface area of about 85,000 km^2, and the distance to the Sun is about 2.7 B km.
 Our beloved Earth is waiting….:

 The Constitution of the World is intended for at least 10,000 years of harmonious living on the happy Terra.
 The Constitution of the World was ready to come into force, and to be put into practice, for the benefit of all people on Terra, on 6 March 2020, and it is ready to remain into force, and enjoyed by all people, at least until 6 March 12020.

MARS: Magnificent!

MOON: Victor Hugo said "All the forces in the world are not so powerful as an idea whose time has come".

MARS: The sooner, the better!

MOON: Yes – in conclusion let's mention an interesting comet: C/2023 A3 (Tsuchinshan–ATLAS) is a comet from the Oort cloud, with a mean diameter of about 3.2 km, discovered by the Purple Mountain Observatory in China on 9 January 2023, and also found by ATLAS South Africa on 22 February 2023. The comet passed perihelion at a distance of 58,343,000 km (0.39 AU) from the Sun, on 27 September 2024, when it became visible to the naked eye.

Orbital period: ≈ 110 millions of years (inbound); ≈ 235,000 years (outbound)

Aphelion (farthest away from the Sun): ≈ 40 trillions km (270,000 AU, inbound); ≈ 568 billions km (3,800 AU, outbound)

Comet nuclear magnitude (M2): 9.2 ± 0.3 (brightest Venus is minus 4.6, Sirius minus 1.46, Alpha Centauri minus 0.1, Vega, +0.03, faintest star to see with naked eye is +7.2 - for bigger numbers, optical astronomical instruments are needed).

Now we wish our readers all the best, and we will continue our dialog very soon.

London - From the Broad Sanctuary, west of St. Margaret's Church, looking southwest to the west part of the north façade (north entrance (left)) of Westminster Abbey (960, 1517, Collegiate Church of St. Peter at Westminster, Anglican abbey hosting daily services, and every coronation since 1066, tower height 69 m, floor area 3,000 m^2).

Bibliography

"The Histories" by Polybius
"Discours de la Méthode" by René Descartes
"Meditationes de prima philosophia" by René Descartes
"Philosophiae Naturalis Principia Mathematica" by Isaac Newton
"Zum ewigen Frieden. Ein philosophischer Entwurf" by Immanuel Kant
Chinese encyclopedia Gujin Tushu Jicheng (Imperial Encyclopedia)
"Encyclopédie" by Jean-Baptiste le Rond d'Alembert and Denis Diderot
"Encyclopaedia Britannica" by over 4,400 contributors
"Encyclopedia Americana" by Francis Lieber
"My Journey at the Nuclear Brink" by William J. Perry
"Breakthrough: Emerging New Thinking" by Martin E. Hellman and Prof. Anatoly Gromyko
"Leading Matters" by John L. Hennessy
"Mathematical Models in International Relations: A Bibliography" by Claudio Cioffi-Revilla
"The Future of the Nuclear Fuel Cycle – An Interdisciplinary MIT Study" by Eugen Shwageraus, George Apostolakis, Paul Hejzlar
"Invent and Wander" by Jeff Bezos
"International Anniversary Anthology" by Kaiti Batalia
Other sources include: United Nations News, UPI, Nature, CNBC, AP, Nasdaq, AAAS, Reuters, EDGAR, AFP, Recode, Europa Press, American Mathematical Society, NDTV India, The Economist, Bloomberg News, Fox News, USA, Deutsche Presse-Agentur, MSNBC, BBC, The Financial Times, Australian Associated Press, Agência Brasil, The Canadian Press (La Presse Canadienne), Middle East News Agency, Baltic News Service, Suomen Tietotoimisto, Athens-Macedonian News Agency, Asian News International, Inter Press Service, Kyodo News, Notimex, Algemeen Nederlands Persbureau, AGERPRES, Newsis, Tidningarnas Telegrambyrå, Swiss Telegraphic Agency, Central News Agency, ANKA news agency, Agenzia Fides

7 November 2023 – OUDIN - TRESOR DES TROIS LANGVES ESPAGNOLE FRANCOISE ET ITALIENNE, GENEVE 1627; OTTARELLI - ITALIAN, ENGLISH AND FRENCH DICTIONARY, 1777; a book about Graece Linguae, 1700 - books from Thomas Jefferson's Library inside the Library of Congress.

Books by Michael M. Dediu

1. Aphorisms and quotations – with examples and explanations
2. Axioms, aphorisms and quotations – with examples and explanations
3. 100 Great Personalities and their Quotations
4. A Great Mathematician and Engineer
5. A Dedicated Engineering Professor
6. Venice (Venezia) – a new perspective. A short presentation with photographs
7. La Serenissima (Venice) - a new photographic perspective. A short presentation with many photos
8. Grand Canal – Venice. A new photographic viewpoint. A short presentation with many photos
9. Piazza San Marco – Venice. A different photographic view. A short presentation with many photos
10. Roma (Rome) - La Città Eterna. A new photographic view. A short presentation with many photos
11. Why is Rome so Fascinating? A short presentation with many photos
12. Rome, Boston and Helsinki. A short photographic presentation
13. Rome and Tokyo – two captivating cities. A short photographic presentation
14. Beautiful Places on Earth – A new photographic presentation
15. From Niagara Falls to Mount Fuji via Rome - A novel photographic presentation
16. From the USA and Canada to Italy and Japan - A fresh photographic presentation
17. Paris – Why So Many Call This City Mon Amour - A lovely photographic presentation
18. The City of Light – Paris (La Ville-Lumière) - A kaleidoscopic photographic presentation
19. Paris (Lutetia Parisiorum) – the romance capital of the world - A kaleidoscopic photographic view
20. Paris and Tokyo – a joyful photographic presentation. With a preamble about the Universe

7 November 2023, detail of Thomas Jefferson Library in the Library of Congress, Washington, DC.

21. From USA to Japan via Canada – A cheerful photographic documentary
22. 200 Wonderful Places, In The Last 50 Years – A personal photographic documentary
23. Must see places in USA and Japan - A kaleidoscopic photographic documentary
24. Grandeurs of the World - A kaleidoscopic photographic documentary
25. Corneliu Leu – writer on the same wavelength as Mark Twain. An American viewpoint
26. From Berkeley to Pompeii via Rome – A kaleidoscopic photographic documentary
27. From America to Europe via Japan - A kaleidoscopic photographic documentary
28. Discover America and Japan - A photographic documentary
29. J. R. Lucas – philosopher on a creative parallel with Plato, An American viewpoint
30. From America to Switzerland via France - A photographic documentary
31. From Bretton Woods to New York via Cape Cod - A photographic documentary
32. Splendid Places on the Atlantic Coast of the U. S. A. - A photographic documentary
33. Fourteen nice Cities on three Continents - A photographic documentary
34. 17 Picturesque Cities on the World Map - A photographic documentary
35. Unforgettable Places from Four Continents, including Trump buildings - A photographic documentary
36. Dediu Newsletter, Volume 1, Number 1, 6 December 2016 – Monthly news, review, comments and suggestions for a better and wiser world
37. Dediu Newsletter, Volume 1, Number 2, 6 January 2017 (available also at www.derc.com).
38. Dediu Newsletter, Volume 1, Number 3, 6 February 2017 (available at www.derc.com).
39. London and Greenwich, - A photographic documentary
40. Dediu Newsletter, Volume 1, Number 4, 6 March 2017 (available also at www.derc.com).

Rome: Accademia Nazionale dei Lincei (1603, the oldest worldwide) has its library in Palazzo Corsini (1740), Via della Lungara 10, Roma.

Rome: Accademia Nazionale dei Lincei (1603) in Villa Farnesina (1510). The author was invited to give a lecture here in 1977.

41. Dediu Newsletter, Volume 1, Number 5, 6 April 2017 (available also at www.derc.com).
42. Dediu Newsletter, Volume 1, Number 6, 6 May 2017 (available also at www.derc.com).
43. Dediu Newsletter, Volume 1, Number 7, 6 June 2017 (available also at www.derc.com).
44. London, Oxford and Cambridge, A photographic documentary
45. Dediu Newsletter, Volume 1, Number 8, 6 July 2017 (available also at www.derc.com).
46. Dediu Newsletter, Volume 1, Number 9, 6 August 2017 (available also at www.derc.com).
47. Dediu Newsletter, Volume 1, Number 10, 6 September 2017 (available also at www.derc.com).
48. Three Great Professors: President Woodrow Wilson, Historian German Arciniegas, and Mathematician Gheorghe Vranceanu – A chronological and photographic documentary
49. Dediu Newsletter, Volume 1, Number 11, 6 October 2017 (available also at www.derc.com).
50. Dediu Newsletter, Volume 1, Number 12, 6 November 2017 (available also at www.derc.com).
51. Dediu Newsletter, Volume 2, Number 1 (13), 6 December 2017 (available also at www.derc.com).
52. Two Great Leaders: Augustus and George Washington - A chronological and photographic documentary
53. Dediu Newsletter, Volume 2, Number 2 (14), 6 January 2018 (available also at www.derc.com).
54. Newton, Benjamin Franklin, and Gauss, A chronological and photographic documentary
55. Dediu Newsletter, Volume 2, Number 3 (15), 6 February 2018 (available also at www.derc.com).
56. 2017: World Top Events, But Many Little Known, A chronological and photographic documentary
57. Dediu Newsletter, Volume 2, Number 4 (16), 6 March 2018 (available also at www.derc.com).
58. Vergilius, Horatius, Ovidius, and Shakespeare - A chronological and photographic documentary.
59. Dediu Newsletter, Volume 2, Number 5 (17), 6 April 2018 (available also at www.derc.com).

7 November 2023 – HICKES – INSTITUTIONES GRAMMATICAE ANGLO-SAXONICAE, OXONIAE 1689; CHIRCHMAIR – Grammatica della Lingua Tedesca, Milano, 1706 - books from Thomas Jefferson's Library inside the Library of Congress.

60. Dediu Newsletter, Volume 2, Number 6 (18), 6 May 2018 (available also at www.derc.com).
61. Vivaldi, Bach, Mozart, and Verdi - A chronological and photographic documentary.
62. Dediu Newsletter, Volume 2, Number 7 (19), 6 June 2018 (available also at www.derc.com).
63. Dediu Newsletter, Volume 2, Number 8 (20), 6 July 2018 (available also at www.derc.com).
64. Dediu Newsletter, Volume 2, Number 9 (21), 6 August 2018 (available also at www.derc.com).
65. World History, a new perspective - A chronological and photographic documentary.
66. World Humor History with over 100 Jokes, a new perspective - A chronological and photographic documentary
67. Dediu Newsletter, Volume 2, Number 10 (22), 6 September 2018 (available also at www.derc.com).
68. Dediu Newsletter, Volume 2, Number 11 (23), 6 October 2018 (available also at www.derc.com).
69. Dediu Newsletter, Volume 2, Number 12 (24), 6 November 2018
70. Da Vinci, Michelangelo, Rembrandt, Rodin - A chronological and photographic documentary
71. Dediu Newsletter, Volume 3, Number 1 (25), 6 December 2018
72. Dediu Newsletter, Volume 3, Number 2 (26), 6 January 2019
73. From Euclid to Edison – revelries in the past 75 years - A chronological and photographic documentary
74. – Socrates to Churchill Aphorisms celebrated after 1960 - A chronological and photographic documentary
75. - Dediu Newsletter, Volume 3, Number 3 (27), 6 February 2019
76. – Hippocrates to Fleming: Medicine History celebrated after 1943 - A chronological and photographic documentary
77. - Dediu Newsletter, Volume 3, Number 4 (28), 6 March 2019
78. - Dediu Newsletter, Volume 3, Number 5 (29), 6 April 2019
79 – Archimedes to Ford: Invention History celebrated after 1943 - A chronological and photographic documentary
80 - Dediu Newsletter, Volume 3, Number 6 (30), 6 May 2019
81 – Sutherland to Pavarotti: Great Singers History - A chronological and photographic documentary
82 - Dediu Newsletter, Volume 3, Number 7 (31), 6 June 2019

7 November 2023, detail of the ceiling of Thomas Jefferson Library in the Library of Congress, Washington, DC.

83 - Dediu Newsletter, Volume 3, Number 8 (32), 6 July 2019
84 – Augustus to Rockefeller: History of the Wealthiest People - A chronological and photographic documentary
85 - Dediu Newsletter, Volume 3, Number 9 (33), 6 August 2019
86 – Pythagoras to Fermi: History of Science - A chronological and photographic documentary
87 - Dediu Newsletter, Volume 3, Number 10 (34), 6 September 2019
88 – Our Future is Sustainable Peace and Prosperity – Moving from conflicts to harmony and peace
89 - Dediu Newsletter, Volume 3, Number 11 (35), 6 October 2019 – World Monthly Report with news
90 – Our Future Depends on Good World Educations – Moving from frail education to solid education
91 - Dediu Newsletter, Volume 3, Number 12 (36), 6 November 2019 – World Monthly Report with News and Suggestions for Sustainable Peace, Freedom and Prosperity
92 – Friendly, Helpful & Smart World Management - Moving from bureaucracy to responsive world management
93 – If You Want Peace, Prepare for Peace! – Moving from preparation for war to preparation for peace
94 - Dediu Newsletter, Volume 4, Number 1 (37), 6 December 2019 – World Monthly Report with News and Suggestions for Sustainable Peace, Freedom and Prosperity
95 – World with One Country & its Ten Friendly Regions - Moving from 195 disagreeing countries, to 1 country with 10 collaborating regions
96 - Dediu Newsletter, Volume 4, Number 2 (38), 6 January 2020 – World Monthly Report with News and Suggestions for Sustainable Peace, Freedom and Prosperity
97 – After 10,000 Years of Conflicts, People want 10,000 Years of Harmony - Moving from continuous wars to stable peace
98 - Dediu Newsletter, Volume 4, Number 3 (39), 6 February 2020 – World Monthly Report with News and Suggestions for Sustainable Peace, Freedom and Prosperity
99 – The Constitution of the World – Moving from many unsustainable constitutions, to just one Constitution of the World

7 November 2023, The Gutenberg Bible in the Library of Congress, Washington, DC.

100 - Dediu Newsletter, Volume 4, Number 4 (40), 6 March 2020 – World Monthly Report with News and Suggestions for Sustainable Peace, Freedom and Prosperity
101 - Dediu Newsletter, Volume 4, Number 5 (41), 6 April 2020 – World Monthly Report
102 - Dediu Newsletter, Volume 4, Number 6 (42), 6 May 2020 – World Monthly Report
103 – World Constitution Implementation – Moving from violent changes, to smooth transition to the Constitution of the World
104 - Dediu Newsletter, Volume 4, Number 7 (43), 6 June 2020 – World Monthly Report
105 - Dediu Newsletter, Volume 4, Number 8 (44), 6 July 2020 – World Monthly Report
106 - It is getting truer and truer – we urgently need the World Constitution: Moving from anarchic changes, to balanced transition to the Constitution of the World
107 - Dediu Newsletter, Volume 4, Number 9 (45), 6 August 2020 – World Monthly Report
108 - World Constitution with Lovely Comments - Moving from many suboptimal constitutions to the much better Constitution of the World
109 - Dediu Newsletter, Volume 4, Number 10 (46), 6 September 2020 – World Monthly Report
110 – World Constitution with Questions & Answers – Moving from many obsolete constitutions to the much better Constitution of the World
111 - Dediu Newsletter, Volume 4, Number 11 (47), 6 October 2020 – World Monthly Report
112 - World Projects - Moving from minor projects to great projects for the World
113 - Dediu Newsletter, Volume 4, Number 12 (48), 6 November 2020 – World Monthly Report
114 - Dediu Newsletter, Volume 5, Number 1 (49), 6 December 2020 – World Monthly Report
115 - World Opportunities for All - Moving from few local jobs, to world opportunities for all
116 - Dediu Newsletter, Volume 5, Number 2 (50), 6 January 2021 – World Monthly Report

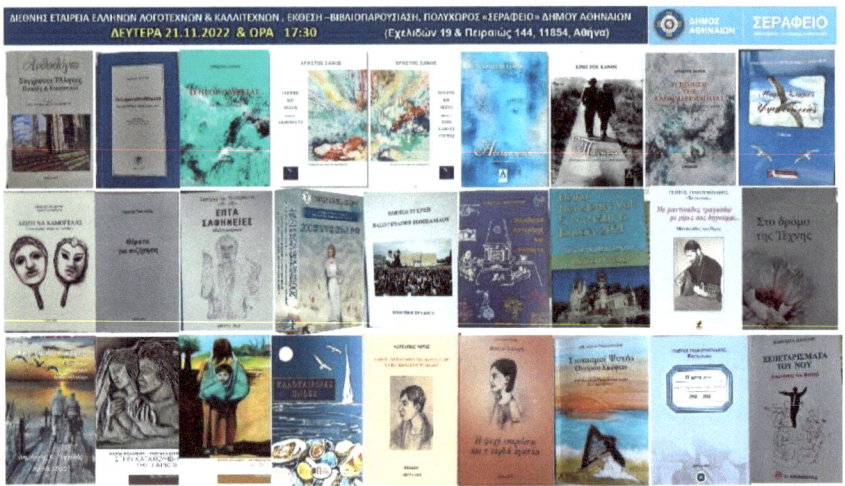

On row 2, the 7th, at an exposition in Athens, Greece, Fall 2022

117 - Self-Managing World - Moving from local ruling top-down, to self-managing world
118 – We are all in the same space boat – Peaceful Terra; Moving from local fragile boats to the solid Peaceful Terra
119 - Dediu Newsletter, Volume 5, Number 3 (51), 6 February 2021 – World Monthly Report
120 - All people ask for Peace + Freedom = Prosperity, Moving from local conflicts to world peace and freedom
121 - Dediu Newsletter, Volume 5, Number 4 (52), 6 March 2021 – World Monthly Report
122 - To pour Peace from a cup full of arms, MELT ALL ARMS! - Moving from arms race, to peace enjoyment
123 - Dediu Newsletter Vol 5, Number 5 (53), 6 April 2021 – World Monthly Report
124 - Bureaucracy is growing like a weed - People want a Quality Change; Yup, that's right! Better life for all!
125 - Dediu Newsletter Vol 5, Number 6 (54), 6 May 2021 - World Monthly Report
126 – What is Life for Homo Sapiens post 2020? – Life is evolution by harmony, not by natural selection for people.
127 - Dediu Newsletter Vol 5, Number 7 (55), 6 June 2021 - World Monthly Report

128 – All Wars and Conflicts are due to HUMAN ERRORS: Moving from perpetual war errors, to friendly collaboration and peace
129 - Dediu Newsletter Vol 5, Number 8 (56), 6 July 2021 - World Monthly Report
130 - WARS + Incompetence = 600 LOST YEARS - Moving from war and incompetence imposed poverty, to peace, harmony and prosperity
131 - Dediu Newsletter Vol 5, Number 9 (57), 6 August 2021 - World Monthly Report
132 – People are something Sacred for People – Moving from wars and incompetence, to peace, harmony and prosperity
133 - Dediu Newsletter Vol 5, Number 10 (58), 6 September 2021 - World Monthly Report
134 – The 71-dimentional Space of Terra's Future: Moving from the war-dimension to peace, harmony and prosperity.
135 – Preventing bad events – Moving from waiting for bad events, to preventing them

On row 2, the 4th, at an exposition in Athens, Greece, Fall 2022

136 - Dediu Newsletter Vol 5, Number 11 (59), 6 October 2021 - World Monthly Report
137 - The 7.8 B people need $7.8 B to change from war to Peace - Moving from trillions for war, to peace, freedom and prosperity

138 - Dediu Newsletter Vol 5, Number 12 (60), 6 November 2021 - World Monthly Report

139 – We can't just shrug off problems – we must solve them! Moving from looking at problems, to solving them

140 - Dediu Newsletter Vol 6, Number 1 (61), 6 December 2021 - World Monthly Report

141 - Today we are here, Tomorrow we are gone - What do we leave to our children and grandchildren? NOT wars, conflicts and poverty, BUT Peace, Harmony and Prosperity!

142 – Self-Motivating World – Moving from local top-down domination, to self-motivating world

143 - Relay Management of the World - moving from conflictual do-undo management, to collaborative world management

144 - Dediu Newsletter Vol 6, Number 2 (62), 6 January 2022 - World Monthly Report

145 - Dediu Newsletter Vol 6, Number 3 (63), 6 February 2022 - World Monthly Report

146 - Regional Projects on Earth, Moving from local and limited projects, to large and important projects

147 - Dediu Newsletter Vol 6, Number 4 (64), 6 March 2022 - World Monthly Report

148 – United & Friendly World Association – Moving from many suboptimal associations to the much better one United & Friendly World

149 - Dediu Newsletter Vol 6, Number 5 (65), 6 April 2022 - World Monthly Report

150 - Our Good Future depends on our Good Present! - moving from ugly conflicts to harmonious peace

151 - Dediu Newsletter Vol 6, Number 6 (66), 6 May 2022 - World Monthly Report

152 – People are at the Center of any Technology – The technology must work for people, not against them

153 - Dediu Newsletter Vol 6, Number 7 (67), 6 June 2022 - World Monthly Report

154 – Harmonious World Managing Orchestration – Moving from a discordant world, to a harmonious world

155 – Preventive Medical Cornucopia – Moving from post-sickness help, to avoiding sickness altogether

7 November 2023 –ARISTOTELIS - Opera omnia, 1597; DALZEL – Sive Collectanea Graeca Minora, Edinburgi, 1787 - books from Thomas Jefferson's Library inside the Library of Congress.

156 - Dediu Newsletter Vol 6, Number 8 (68), 6 July 2022 - World Monthly Report
157 - Metastability of the World - Moving from chronic instability to unshakeable global stability
158 - Dediu Newsletter Vol 6, Number 9 (69), 6 August 2022 - World Monthly Report
159 – Earth is one Big Playground for all Children – Moving from war devastated Earth, to a peaceful playground for all children
160 - Dediu Newsletter Vol 6, Number 10 (70), 6 September 2022 - World Monthly Report
161- World Harmonious System: Enjoy the thrill of being on a Great Planet - Moving from many obsolete ruling systems, to the much better World Harmonious System
162 - Dediu Newsletter Vol 6, Number 11 (71), 6 October 2022 - World Monthly Report
163 - Dediu Newsletter Vol 6, Number 12 (72), 6 November 2022 - World Monthly Report
164 – Invisible Movements in the thinking of People – moving from obsolete thinking, to future-oriented thinking
165 - Socrates: So, we have 8 B people on Earth, Plato, where do we go from here? Moving from war and poverty, to peace, freedom & prosperity
166 - Dediu Newsletter Vol 7, Number 1 (73), 6 December 2022 - World Monthly Report
167 – Zero errors management – Moving from errors after errors, to errors-free management
168 - Dediu Newsletter Vol 7, Number 2 (74), 6 January 2023 - World Monthly Report
169 – People find joy in helping people – Moving from frustrations and worries, to the joy of helping people
170 - Dediu Newsletter Vol 7, Number 3 (75), 6 February 2023 - World Monthly Report
171 – People's Fundamental necessities - Moving from ignoring people's fundamental necessities, to implementing all of them
172 - Dediu Newsletter Vol 7, Number 4 (76), 6 March 2023 - World Monthly Report
173 - Dediu Newsletter Vol 7, Number 5 (77), 6 April 2023 - World Monthly Report

7 November 2023 – A view of the interior of the Library of Congress.

174 – 2123 A Wonderful Year! – Moving from errors, to a much better future
175 – 2222 A Splendid Year! – Moving from 10 negative responses, to 10 positive responses
176 - Dediu Newsletter Vol 7, Number 6 (78), 6 May 2023 - World Monthly Report
177 – 2333 An Exemplar Year! – Moving from slow and incorrect, to faster and better projects
178 - Dediu Newsletter Vol 7, Number 7 (79), 6 June 2023 - World Monthly Report
179 - Dediu Newsletter Vol 7, Number 8 (80), 6 July 2023 - World Monthly Report
180 – 2444 A better Management Year! Moving from a dysfunctional management to a good results world management
181 - Dediu Newsletter Vol 7, Number 9 (81), 6 August 2023 - World Monthly Report
182 - Optimal Aging in the Medical Year 2555 - Moving from aging-as-you-go, to optimal aging in 2555
183 - Dediu Newsletter Vol 7, Number 10 (82), 6 September 2023 - World Monthly Report
184 - Dediu Newsletter Vol 7, Number 11 (83), 6 October 2023 - World Monthly Report
185. – The Great Musical Year 2666 – Moving from sporadic music activity to a really good musical year 2666
186 - Dediu Newsletter Vol 7, Number 12 (84), 6 November 2023 - World Monthly Report
187 – Good Humor in the Joyful Year 2743 – Moving from little humor, to great humor in the joyful year 2743
188 - Dediu Newsletter Vol 8, Number 1 (85), 6 December 2023 - World Monthly Report
189 – From the lovely Year 2823 Admiring photos from 2023 – Moving from nice photos in 2023, to a review of them in 2823
190 - Dediu Newsletter Vol 8, Number 2 (86), 6 January 2024 - World Monthly Report
191 – History Events reviewed from Year 2943 – Moving from shy history, to historia magistra vitae est in Year 2943
192 - Dediu Newsletter Vol 8, Number 3 (87), 6 February 2024 - World Monthly Report

193 – Robust Cognitive Vitality in Transformative Year 3000: Moving from a virus infected cognitivity to virus free cognitive vitality in Year 3000

194 - Dediu Newsletter Vol 8, Number 4 (88), 6 March2 024 - World Monthly Report

195 – Brain's Cognitive Power in the Improvement Year 3001 – Moving from slow cognitivity, to improved cognitive function and speed in Year 3001

196 - Dediu Newsletter Vol 8, Number 5 (89), 6 April 2024 - World Monthly Report

197 – Rejuvenating Immunity in Elderly People in the Refreshing Year 3024 - Moving from reduced immunity in seniors, to more youthful and effective immunity

198 - Dediu Newsletter Vol 8, Number 6 (90), 6 May 2024 - World Monthly Report

199 - Immanuel Kant's Perpetual Peace updated and implemented by Year 3124 - Moving from philosophical Perpetual Peace, to fully updated and implemented Perpetual Peace

200 - Dediu Newsletter Vol 8, Number 7 (91), 6 June 2024 - World Monthly Report

201 – Predictive technology in the Technological Year 3243 – Moving from a slow technology advance, to a rapid predictive technology by Year 3243

202 - Dediu Newsletter Vol 8, Number 8 (92), 6 July 2024 - World Monthly Report

203 - Advanced Culinary Medicine in the Culinary Year 3333 - Moving from an incipient phase, to an advanced culinary medicine in the Year 3333

204 - Dediu Newsletter Vol 8, Number 9 (93), 6 August 2024 - World Monthly Report

205 – Delightful Olympiads in the Olympic Year 3424 – Moving from isolated Olympics, to an exciting World Olympics in 3424

206 - Dediu Newsletter Vol 8, Number 10 (94), 6 September 2024 - World Monthly Report

206 - Dediu Newsletter Vol 8, Number 11 (95), 6 October 2024 - World Monthly Report

Italy, Venezia - The south of La Piazzetta, the south of Piazza San Marco, with gondole, and wedding pictures of a Japanese couple.

www.ingramcontent.com/pod-product-compliance
Lightning Source LLC
Chambersburg PA
CBHW040215220526
45473CB00001B/1